Statistics and Computing

D0375595

Series Editors:
J. Chambers
W. Eddy
W. Härdle
S. Sheather
L. Tierney

Springer
New York
Berlin
Heidelberg
Barcelona
Budapest
Hong Kong
London
Milan
Paris
Santa Clara
Singapore
Tokyo

Statistics and Computing

Andreas Krause
Melvin Olson

The Basics of
S and S-Plus

With 34 Illustrations

 Springer

Andreas Krause
GeneData AG
Postfach 254
4016 Basel
Switzerland
e-mail:
andreas.krause@genedata.com

Melvin Olson
Ares-Serono
15bis, Chemin des Mines
Case Postale 54
1211 Geneva 20
Switzerland
e-mail:
s_s_olson@compuserve.com

Series Editors:

J. Chambers
Bell Labs, Lucent Technologies
Murray Hill, NJ 07974
USA

W. Eddy
Department of Statistics
Carnegie Mellon University
Pittsburgh, PA 15213
USA

W. Härdle
Institut für Statistik und Ökonometrie
Humboldt-Universität zu Berlin
D-10178 Berlin
Germany

S. Sheather
Australian Graduate School
  of Management
Kensington, New South Wales 2033
Australia

L. Tierney
School of Statistics
University of Minnesota
Minneapolis, MN 55455
USA

Library of Congress Cataloging-in-Publication Data
Krause, Andreas.
    The basics of S and S-Plus / Andreas Krause, Melvin Olson.
        p.   cm. — (Statistics and Computing)
    Includes index.
    ISBN 0-387-94985-2 (softcover : alk. paper)
    1. S-Plus.   2. Mathematical statistics — Data processing.
    I. Olson, Melvin.   II. Title.   III. Series.
    QA276.4.K73   1997
    519.5′0285′53 — dc21   97-10382

Printed on acid-free paper.

Production managed by Victoria Evaretta; manufacturing supervised by Jacqui Ashri.
Photocomposed pages prepared from the authors' LaTeX files.
Printed and bound by R.R. Donnelley and Sons, Harrisonburg, VA.
Printed in the United States of America.

9 8 7 6 5 4 3 2 1

ISBN 0-387-94985-2 Springer-Verlag New York Berlin Heidelberg   SPIN 10557457

Preface

S-PLUS is a powerful tool for interactive data examination, creation of graphs, and implementation of customized routines based on the S language of AT&T Bell Labs (Lucent Technologies). The modern concept together with its flexibility makes it appealing to data analysts from all scientific fields. Among its many strengths are its powerful but flexible graphics routines, and its facility for writing functions whereby you may either modify an existing function, or create one of your own. Most standard statistical techniques and many non-standard ones besides are already programmed in S-PLUS, making exploratory and formal statistical analysis very easy.

Our book was written to give a solid but quick introduction to the S-PLUS environment. If you have never used S-PLUS before, this book will get you up to speed quickly, as it is full of examples and insightful remarks on how S-PLUS works. Nevertheless, we start from first principles and make few assumptions about prior knowledge of computing and statistics.

Although our book is primarily intended for the S-PLUS novice, the material covered extends to more advanced topics and contains many hints beneficial to those people who already have some knowledge of S-PLUS or are not aware of the many enhancements that have been made to the system over the past several years.

The structure of our book, with its detailed exercises and solutions at the end of each chapter, reflects its origin from lecture notes and homework exercises. This is an obvious way to use the book in the future. However, the exercises were kept as an integral part of the book to encourage active thinking on the part of the reader, as opposed to the more passive learning obtained by merely reading it. Comparison of your solutions to ours is strongly encouraged, as there is no single or correct approach to most problems.

The best way to use the book is interactively. That is, you should read the book while seated at your computer and repeat all of the examples as you read them. For getting an overview, reading all the text material should take no more than a day or two, while working through the examples will take some additional days. If used as a basis for lectures, typically one chapter forms the material for a 90-minute lecture, and the exercises repeat and reinforce the information contained in that chapter.

Using this book to lay the foundation, the reader is thus enabled to use S-PLUS with all its basic capabilities. With the groundwork in place, it is then possible to tackle the really sophisticated functions and to write your own. If a specific problem is not covered in the book, you should be able to find the answer in one of the voluminous manuals on S-PLUS.

This book has undergone many revisions and reflects not only our own experiences, but also those of the many people who influenced it by giving comments and proposing alternative ways to approach the subject. Students of the statistics department at the University of Dortmund, as well as students of the economics department of the University of Basel, were most helpful.

Many more people influenced the various stages of this manuscript and contributed directly or indirectly. They are far too numerous to mention. We would like to thank all of them. Special thanks go to Axel Benner (Heidelberg), Tom Cook (Wisconsin), Peter Hartmann (St. Gallen), Thomas T. Kötter (Berlin), and Tony Rossini (South Carolina).

We also owe a big thanks to the S-PLUS developers and distributors. Doug Martin, Brian Clark, and Charlie Roosen at the Data Analysis Products Division of MathSoft, Inc. (formerly Statistical Sciences) in Seattle, and Reinhard Sy at GraS (Berlin), provided generous support and an early beta version of the latest S-PLUS release.

The editors at Springer-Verlag (New York), John Kimmel and Martin Gilchrist, provided very helpful comments during a continous good cooperation.

And, last but not least, it is important to mention that this book would not have been possible without the tolerance and support of our wives, Michaela Jahn and Sabine Olson. We dedicate this book to them.

Basel and Geneva   Andreas Krause and Melvin Olson
May 1997

Contents

List of Figures

List of Tables

1. Introduction

Over the years, the S language and S-PLUS have undergone many changes. Since its development in the mid-seventies, the three main authors of S, Rick Becker, John Chambers, and Allan Wilks, have enhanced the entire language considerably. All their work was done at Bell Labs with the original goal of defining a language to make it easier to do repetitive tasks in data analysis like calculating a linear model.

In the following years, many people contributed to the S project in one form or another. People outside Bell Labs also became aware of the interesting development and took part in it, and this is to a great extent the way S and S-PLUS are still developed today. A very lively user community works with and on S/S-PLUS, and they especially appreciate the S style of working in an integrated environment. Special strengths are the extremely modern and flexible language, which has many elements of an interactive C, Lisp, and APL, and good graphics capabilities.

It is noteworthy that the authors do not consider S as a primarily statistical system. The system was developed to be flexible and interactive, especially designed for doing data analysis of an exploratory nature, which began to boom in the late seventies after the release of Tukey's book (1977) on the subject. Most of the statistical functionality was added later, and many statistics routines like estimation, regression, and testing were incorporated by the S-PLUS team.

This chapter describes the development of S and S-PLUS over the years, clarifies the differences between the two, and points to some further references.

1.1 The History of S and S–PLUS

The S Language, and some years later the NEW S, were developed at AT&T Bell Labs in the late seventies and early eighties, mainly by Rick Becker and John Chambers. Some years later, Allan Wilks joined the core team. Since then, several other people have been involved in the project. Becker (1994) describes in great detail the foundation and development of S, and points out some future directions.

There is an electronic archive of papers written at AT&T, many of them being reports on or applications in S. This archive is accessible via the World Wide Web (WWW) under

http://netlib.att.com/cm/ms/departments/sia/doc/index.html.

John Chambers' home page refers to further documents and can be accessed via http://cm.bell-labs.com/cm/ms/departments/sia/jmc.

The year 1976 can perhaps be viewed as the founding year of S. The first concepts were developed and implemented, along with extended discussions. At first, the "system" consisted of a library of routines together with an interface, such that the kernel code itself could be kept unmodified. In 1981, the S team decided to rewrite the system in C and port it to the UNIX operating system. Since 1981, the source code has been available for interested people outside Bell Labs.

The next years revealed a strong increasing interest amongst statisticians and data analysts in using the system, which was called S by then. It is remarkable that the important steps in the development of S were all marked by books, such that S users today talk about the *Brown Book* Era, the *Blue Book* Era, and the *White Book* Era.

In 1984, as the interest in S began to rise, a real manual was necessary. The first book, today referred to as the *Brown Book*, was written by Becker and Chambers (1984). This version of S is now referred to as "Old S," as no version numbers existed at the time.

The QPE (Quantitative Programming Environment) developed by John Chambers set a milestone in the development of S. In 1988, it introduced the function concept (replacing the former macros), and new programming concepts were added. This work is described in the *Blue Book* (Becker, Chambers, and Wilks, 1988).

During all these years, the user community added substantial functionality to S, and many sophisticated techniques like tree regression, nonparametric smoothing, survival analysis, object-oriented programming, and new model formulation became a part of S. This step in the development was manifested and accompanied by the *White Book* (Chambers and Hastie, 1992).

Chambers (1995) gives an outlook on the new developments of S, which will comprise an extended and unified object-oriented approach, event management facilities, a pre-compiler for semantic analysis, and an online documentation concept (every object carries its own documentation).

In 1987, Douglas Martin at the University of Washington, Seattle, founded a small company to make S more popular. He realized that the major drawback of S was the need of professional support for end-users. Hence, he started the company Statistical Sciences, Inc. (StatSci), a division of Math-Soft, Inc., since 1994. StatSci added more functionality to S, ported it to other hardware platforms, and provided the necessary support for technical and statistical problems. The enhanced version of S received a new name: S-PLUS. S-PLUS helped popularize S among nontechnical people. StatSci also ported S-Plus to the only non-UNIX platform, releasing S-Plus for DOS in 1989 and S-Plus for Windows in 1993.

Version 4 of S-Plus has undergone a major design revision. The graphical user interface (GUI) was changed to adopt the Windows standard under Windows. Much of the functionality is available via menus and buttons, and the graphics are shown in a graph sheet. The elements of the graph can be edited and the graph sheet itself can be saved.

S is still the heart of the system, and the core S team continues to work on the S system. The whole S functionality is incorporated in S-Plus and enhanced, and today the S system is no longer publically available. In the remainder of the book, we will use S-Plus as the standard reference.

Further Reading about S-PLUS

In recent years, several publications related to S-Plus have appeared, all with their special topics. The following gives a short overview.

The fundamental *Blue Book* (Becker, Chambers, and Wilks, 1988), is always worth looking at. It contains a very easy-to-read introduction to the principal ideas of the whole system and has good examples.

If you are interested in more advanced statistical techniques like nonparametric smoothing and high-dimensional techniques, Chambers and Hastie (1992) and Härdle (1991) give an introduction to these topics, along with applications in S.

Spector (1994) gives a comprehensive overview by including more advanced material, making it a good reference book.

The latest book by Venables and Ripley (1997) is a comprehensive introduction to statistical modeling techniques in general, and might serve as good further reading for advanced topics in statistics, as well as a general reference book. In the same direction is the German book by Süselbeck (1993).

S-Plus itself comes with rich documentation. In addition to the standard installation, user's and reference manuals, there is a programmer's manual reflecting some topics especially suited for experienced programmers in S-Plus and programmers in other languages. Then there are two very compact introductory books, the *Gentle Introduction to* S-Plus and the *Crash Course in S-PLUS*, which give short overviews of S-Plus functionality.

If you are interested in learning more about statistical methods, together with their theoretical background and application in S-Plus, the user manuals do a very good job at introducing the methodology by going through the available functionality in S-Plus.

Finally, it is definitely worth subscribing to S-News, an electronic newsgroup which is further discussed in Section 11.3.

1.2 S-PLUS on Different Operating Systems

Sometimes it is important to know about differences in software on various
hardware systems, or under different operating systems, respectively. This
can be the case if you work on more than one computer system with S-PLUS
(and therefore need to exchange data files), or if you want to be informed
about the differences before deciding in favor of a specific system. In this
section, we discuss some details about different systems supported by S-
PLUS. In addition, the chapter provides some basic information about the
general setup of files and structures in S-PLUS. More information on the
S-PLUS internal workings can be found in Section 10.1.

At present, S-PLUS supports two major operating systems, UNIX (with
most of its variants) and DOS/Windows. We concentrate on machines that
are still supported and omit hardware platforms like Convex, NeXT, and
DOS (without Windows), for which only older versions are available. For the
Macintosh, refer to Section 11.5 (page 230) about the R system.

1.2.1 UNIX

In general, all standard UNIX platforms are supported by S-PLUS. Table 1.1
summarizes the currently supported hardware and operating systems.

Table 1.1. UNIX systems supported by S-PLUS

Manufacturer	Hardware System	Operating System
Digital Equipment (DEC)	DECstation	Ultrix
	Alpha	OSF
Hewlett Packard	9000 series	HP-UX
IBM	RS/6000	AIX
Silicon Graphics	Iris 4D, Indigo	IRIX
SUN	SPARC	Solaris, SUN OS

As the S source code is no longer available, machines not binary compatible
to the ones supported are not able to run S or S-PLUS, respectively.

S-PLUS has minimum requirement specifications regarding main memory and hard disk size. In general, if S-PLUS fits on a hard disk (about 50
megabytes are required), it will also run on the machine. As a side note,
S-PLUS consumes and releases memory dynamically during a session, depending on the needs. Therefore, it does not run out of memory until there
is no more main memory and swap space available. If S-PLUS runs out of
main memory (RAM), the operating system assigns virtual memory, that
is, hard disk space, as a substitute. As this slows down the execution time
dramatically, the machine should be equipped with a reasonable amount of

memory, say at least 16 MB. For improving performance, main memory is the first speedup factor. If you are not satisfied with the performance, watch for permanent hard disk access while executing commands, or use a monitoring tool to track swapping activity because of lack of memory.

1.2.2 DOS/Windows

Since 1989, S-PLUS also runs on IBM-compatible Personal Computers (PCs) under the DOS system. It was later rewritten to run under MS Windows, which is currently the only PC version maintained.

S-PLUS on an IBM compatible PC needs about 40 MB of hard disk space in its standard installation, and a few more including all options. For standard purposes, 16 MB of RAM will suffice, but in the long term increasing the size should be considered.

1.2.3 Data Transfer

S-PLUS stores all data created during a session in a data directory. If data is given in text files, the transfer is no problem. If transfer from the data directories is desired, the general way is to dump all data to a text file (this is a single command in S-PLUS) and restore it on the target machine. (The functions to use are `dump` and `restore`.) Some of the supported machines even have compatible storage format so that copying the physical file from one data directory to the other is sufficient.

There are also other tools available. Reading in standard formats like Excel, dBase, and SAS can be done automatically.

1.2.4 Implementation Differences

S-PLUS has some differences in its implementation between the UNIX and the DOS/Windows versions.

Since DOS does not support file names with more than eight characters, S-PLUS has a built-in mechanism for storing all variables with longer names in files named __n, where n is a number. There is a file named ___nonfi, where the given variable names and the corresponding file names are stored. The user sees only the regular mechanism from within S-PLUS and does not need to worry about this. The same applies to the .Data directory under UNIX, which has become _Data under DOS/Windows.

1.3 Notational Conventions

By now you must be eager to get started, but it might be worth reviewing these few notational conventions first, as they are used throughout the book.

To begin with, you must be aware that when running S-PLUS, you will be asked for a new command with the greater than sign: >. We use the > in the book to denote the S-PLUS prompt. Also, if a single command extends over one line of input, the prompt changes to the plus sign: +. A preview of this is shown below.

```
> This is where an S-PLUS command would appear
+ and notice that the prompt changes on the second line.
```

If a longer listing or function is shown, the prompt is typically omitted, as we do not assume interactive input, and reading the code is easier.

There are occasional examples of commands to either the UNIX or DOS shells. For these examples, no prompt is used.

All commands in S-PLUS are actually calls to functions. To highlight the occurrence of a function in S-PLUS, or one of the parameters to it, we have written them in a special font as with the example, `print`.

When presenting commands, we sometimes include descriptive text. These descriptions are separated from the actual commands by the number sign #, according to the S-PLUS syntax.

A summary of these conventions is found in Table 1.2.

Table 1.2. Notational conventions

Convention	Explanation
>	S-PLUS prompt
+	Command has continued onto next line
Commands	Are in typewriter font
No prompt	For calls to UNIX or DOS shell
#	Separates comments from S-PLUS commands
placeholders	Are in italic. You need to replace them by an appropriate expression, like *filename* needs to be replaced by a valid file name
[MENU]	Menu entries and buttons are referred to in this font
Note	Notes point out something important like a practical example, an application, or an exception

2. System Design

The general layout of the S-PLUS system is similar to many popular windows systems in that it has pull-down menus at the top and toolbars just below the menus. To use such a system it is useful to be a little familiar with basic point-and-click operations and how to use a mouse.

For those who are not that comfortable with window- and icon-based software, the following paragraphs provide a crash introduction to the essentials. The pull-down menus across the top are used to group categories of commands or options to be set. Under the [FILE] menu we find actions relating to files ([OPEN], [CLOSE], [IMPORT], [SAVE]) as well as to exiting the system. The toolbars below the menus contain buttons that are convenient shortcuts to commands found through the layers in the menus. Some of the toolbar buttons (e.g. [PLOTS2D]) open a palette containing a myriad of options to complete your task.

Using a mouse efficiently is important to get the most out of the system. Clicking once on the left mouse button is usually used to highlight an item in a list (e.g. a file out of a list of files) or to select a menu heading or button from a toolbar. You will not always be able to guess at the function of a toolbar button merely by looking at the icon, but if you are at a loss, simply position the mouse over the button in question and a short text description will appear below it. Double-clicking on the left mouse button is used to select and execute. Examples include double-clicking on a file name to select it and start the function, or on a part of a graph to select and edit it. After having selected an item by clicking once with the left mouse button, the right mouse button opens a context menu, which changes depending on the item selected.

2.1 Windows Components

When you open S-PLUS - under UNIX by entering

```
Splus
```

and under Windows by double-clicking on the S-PLUS icon - you are greeted by a screen layout as shown in Figure 2.1. This may vary slightly according to the version of S-PLUS you are using. The main elements that are visible include the Object Browser, the Command Window, the menus, and the toolbar. Optionally, a graphics window can be opened. A short description of each of these components is given in the next several sections.[1]

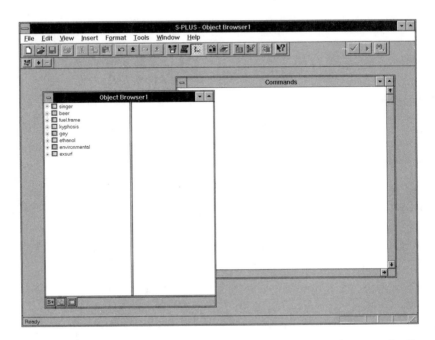

Figure 2.1. The S-Plus screen and its components, the Object Browser, the Command Window, and the toolbars.

2.1.1 Object Browser

The Object Browser is used to get an overview of what is available on the system including data, functions, and graphs. The tabs at the bottom of the window allow you to choose between the categories of objects just mentioned. The Object Browser can be opened either from the menu or from a toolbar button. Not only can it show what is available, but with data, for example, the Object Browser is used to view (browse) and even edit data. Select the data you want to edit and click on the right mouse button to open a dialog box. If you select [OPEN VIEW] from the dialog box, a spreadsheet will open containing the selected data. Once the spreadsheet is open, the data can be edited.

---

[1] The Object Browser, the toolbars, and some other components are not present if S-Plus Version 3 is run.

2.1.2 Command Window

The Command Window is actually the heart of the S-PLUS system. Every command that is performed via menus and buttons can be issued as a command directly from the prompt in the Command Window. In addition, there are many functions that can only be run in the Command Window. As examples, the functions of programming, personalized functions, data subsets, and logical operations are still only available through the use of the Command Window. Its other main use is to show the output from statistical commands entered through the menus or toolbar buttons.

2.1.3 Toolbars

The main toolbar contains many familiar commands including [OPEN], [SAVE], [PRINT], [COPY], [PASTE], and [UNDO]. In S-PLUS, however, one may also open the Command Window and Object Browser, open a new Graph Sheet, open the History Log, and create 2D and 3D graphs.

An additional feature of the toolbars in S-PLUS is that they are context-sensitive (smart). Open a Graph Sheet, for example, and extra buttons will appear in the toolbar area that are specific to Graph Sheets.

2.1.4 Graph Sheets

Graphs are drawn in windows referred to as Graph Sheets. Starting with Version 4.0, components of graphs can be edited and redefined by double-clicking on the component of interest. Labels can be changed, colors modified, axes redefined, and more, all through menus and dialog boxes available simply by clicking in the Graph Sheet. The [INSERT] menu is useful to add features and components to graphs which already exist in a Graph Sheet.

2.2 Working with Menus and Buttons

If you want to use S-PLUS having had no introduction, use the menus and toolbar buttons. For the most part, you should find them to be self-explanatory. However, we have described a few key functions in a bit more detail to get you on your way.

2.2.1 Importing Data

You probably have your own data that you want to analyze, so the first thing you have to know is how to import it into S-PLUS. There is an [IMPORT DATA] facility located in the [FILE] menu. In the [IMPORT DATA] facility, you will be asked for the name you want for your data (which may be more than 8 characters) and have the usual Windows boxes for specifying the name and location of your data file. Pay careful attention to the type of data ([TYPE] menu) that you have and properly define it in the corresponding box. The data file types available are shown in Table 2.1 below.

Table 2.1. File types available with the import facility

File Format			
ASCII	ASCII format	dBase	Excel
Gauss	Lotus 1-2-3	Paradox	Quattro Pro
SAS	SAS Transport	Sigma Plot	S-PLUS
SPSS	SPSS Export	STATA	Systat

This list has been extended from earlier versions, most notably by the addition of SAS files. Users of earlier versions of S-PLUS may use the `sas.get` function available from the Statlib archive (see Section 11.4, page 230).

When the spreadsheet containing newly imported data is closed, the new data automatically appears as a new entry in the Object Browser.

2.2.2 Graphs

Graphs are created by connecting the data shown in the Object Browser to a graphical function represented by a button in a palette. In the Object Browser, the data is displayed with its full name, and after a click on the data object's name, the elements of the selected data set are displayed in the righthand side of the window. By clicking on these elements, the set of variables to be plotted can be selected. To select a second and third variable, hold down the `<Ctrl>` key and click on the element.

Before actually displaying the data graphically, the graph palettes need to be opened, if they are not open yet. Figure 2.2 shows a screen together with the palettes for 2D and 3D graphics. Open the 2D and 3D graph palettes, if they are not yet open, by using the icons in the top menu bar, or by clicking on the [TOOLBARS] selection in the [VIEW] menu.

Once the palette is open, you might want to move the mouse over the different icons slowly. If the mouse stops for a second, the name of the method represented by the button is displayed. If the data has been selected, select the method of displaying the data by clicking on it. S-PLUS executes the graphical method right away by using the selected data.

Note | For creating a graph, the order of selection of variables is important. The variable selected first becomes the x-variable, the second variable selected becomes the y-variable, and if a third variable is selected, it becomes the z-variable.   ◁

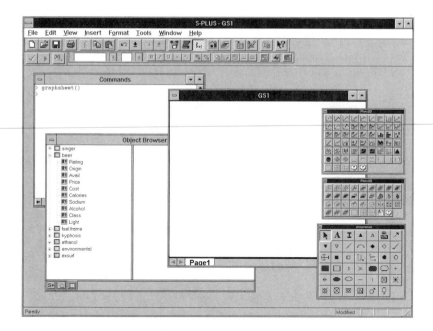

Figure 2.2. The S-PLUS screen with graphics window and palettes.

Once data have been selected, another graphical method can be applied by simply clicking on a different graph button.

Note If you click on a graphics button and nothing happens, this probably means that the data selected and the graph type chosen are not compatible. Selecting two variables and clicking on a 3D graphics method is an example of such a situation.   ◁

Once a basic graph has been created, it can be edited. Double-click on the element to modify, for example, the axis. A window opens and offers the possibility of modifying all the components like range, tickmarks, color, width, label, and much more.

Using the annotate palette, further elements such as text or graphics symbols like rectangles and arrows can be inserted into the graph.

2.2.3 Data and Statistics

Data can be summarized by using graphical display techniques, but also by statistics like the mean, median, minimum, and maximum values.

There are [DATA] and [STATISTICS] menus in the top menu bar that contain a large set of routines for retrieving information about a data set ([DATA]) and routines to process the data ([STATISTICS]).

First select the method to use, then specify the data set and the options you want in the menu fields. The method is applied to the data by clicking on the [OK] button. The result is displayed in the Command Window, together with the function call to an internal S-PLUS function.

In this way, data can be summarized by looking at minimum and maximum values, at quantiles, or at the correlation between variables. Data can be processed by carrying out a t-test, a regression model, or any other method, and toggle buttons can be set on or off to choose options like paired or unpaired or the degree of a polynomial to fit.

The command window informs you about the function called, plus the selected parameters and options. We will learn more about functions later on. For the moment, it might be of interest that this function can be called again with maybe slightly modified settings, without using the menu interface. This is helpful if a series of routines should be applied to many different data sets.

2.2.4 Customizing the Toolbars

The toolbars or palettes can be customized to some extent. It is possible to select which palettes show up on the screen. Select [VIEW] from the top menu and click on [TOOLBARS]. A window opens which lets you choose which toolbars should be shown. Clicking on any of the selection fields pops up or removes the corresponding palette immediately on the desktop. These settings are stored and reused when S-PLUS is started again.

The shapes of the palettes might not be satisfactory, depending on your screen resolution, personal preferences, and more. Clicking on one of the four edges allows you to resize the palettes, which is automatically set such that all icons fit.

Furthermore, the palette buttons can be displayed in palette form or in menu form. Click on the palette and drag it to the top where the primary button bar is located. Drop it and the palette becomes a horizontal menu bar. Dragging the bar away from the top recreates a palette.

Adding new palettes and buttons

The palettes and menu buttons are open for extension. You can add menu buttons to an existing palette and create new palettes. You can generate new toolbars via the [VIEW] menu by selecting the [TOOLBARS] entry and clicking on [NEW]. This produces a new palette on the screen, containing a single button.

You can add a new button by selecting the toolbar and opening the context menu with the right mouse button. The entry [NEW BUTTON] opens up a window where you can specify the name of the button, the text to show when the mouse is over the button, and the S-PLUS function to carry out. (You will learn more about functions as you proceed.)

You can modify an existing button by selecting the button and opening its context menu (right mouse click). A property dialog displays the current settings.

2.3 Learning the System

The graphical user interface (GUI) you just looked at is brand new. S-PLUS has incorporated most of the drag and drop philosophy of modern software. As mentioned before, the origins of S-PLUS lie in the late seventies and early eighties. S-PLUS is very popular because of its ability to process data in a flexible language. The modern concept and the easy-to-use library are still the core of the system and the whole power is in a way hidden behind the simple prompt in the Command Window. Only by learning more about the entire language and its concepts you will get an idea of the possibilities S-PLUS has to offer for data analysis.

The remainder of the book will primarily focus on the language S-PLUS. You will complete several sessions of analyzing data and encounter a variety of statistics routines along the way. Only some of these routines are available through the GUI. The following tour will uncover the full power of the system.

3. A First Session

To begin the book, we introduce the most basic commands so that you can get started with S-PLUS and get a feeling for how the system operates. The commands and principles discussed here may be the most basic, but they are also the most important and the ones that are used the most often. The chapter is designed to cover enough material for your first session with S-PLUS. You will be surprised at how much you can do after this chapter.

Our first session begins with some general information on using S-PLUS, including how to start and quit, and how to access the help system. This is followed by some useful commands and how to do basic arithmetic. Logical variables are covered and the chapter ends with a review of the material from the chapter followed by hands-on exercises.

To greatly enhance your learning of S-PLUS, you should be seated at your computer while completing the exercises and also while reading each chapter.

3.1 General Information

We begin with general information on how to start and quit the system, and how to access the online help facility.

3.1.1 Starting and Quitting

The very first thing you need to know is how to start S-PLUS. If you have S-PLUS on Windows, all you have to do is double-click on the S-PLUS icon and a new S-PLUS window will be opened that should look approximately like the one in Figure 2.1 (page 8). If you're running S-PLUS under UNIX, simply enter Splus at the command prompt. The S-PLUS command prompt (>) appears and indicates that S-PLUS is ready to accept your commands.

Now that you have started S-PLUS, you will, at some point, want to quit. You can quit S-PLUS by double-clicking on the upper-lefthand corner of the window or clicking once on [FILE] and once on [EXIT].

3.1.2 The Help System

If you get stuck, don't panic. S-PLUS has an online help system. To start the help system, simply type `help()` at the S-PLUS command prompt. The help system is designed such that you do not have to know exactly what you're looking for when you access it. It has a table of contents with commands categorized by function, and a search system that operates by typing in the command or subject on which you want information. Based on the subject, rather than the specific command, you can get information on your topic of interest. Of course if you know the name of a specific S-PLUS command, entering `help(`*commandname*`)` will open a new help window with instructions on the functionality of *commandname*. If you are using Windows, you can always click on [HELP] in the menu bar (or hit the <F1> key). If you are using UNIX, you can start a separate help window by entering `help.start()`.

The `help(`*commandname*`)` form of accessing the help system is especially useful when you know the name of a command but have forgotten the specific options it has, the input it requires, or the output it produces. The general `help()` command is useful if you don't know an actual S-PLUS command name, but only a particular subject.

If you prefer to use the mouse, the help system can as well be started by clicking on the [HELP] menu or on the separate help icon. This enables also to use the help system without actually starting the entire system.

3.1.3 Before Beginning

We already covered notational conventions in the introductory chapter. Here we present a few hints to keep in mind before you continue, a couple of which repeat the notational conventions.

We present all S-PLUS commands in `typewriter` font and they may appear either within the body of the text, as in `help()`, or on a separate line with the S-PLUS command prompt, as with

```
> help()
```

All commands in S-PLUS must contain opening and closing parentheses. Options to the commands are quite often included between the parentheses to further specify how a command should be used. The parentheses delimit the command and signify that an execution is to be made. If the parentheses are omitted, the contents of the variable (if data) or the definition of the command will be printed. Try typing "`help`" with no parentheses and the internal commands associated with it are printed on the screen.

It will sometimes be necessary to add a comment to either an S-PLUS command or the output from one. The comment will be denoted with the number sign (`#`).

The act of assigning a value or a function to a variable name occurs all the time in S-PLUS, and it is important that you are familiar with it. In mathematics, you would write, x=3. In S-PLUS, you write x <- 3, where the "<-" is a combination of the less than (<) and minus (-) signs. Optionally, the underscore (_) may be used, as in x_3. Throughout the book we use "<-" because it is easier to read, but in practice you may prefer the underscore because it is quicker to type.

Note This convention of writing notes is to catch your eye and draw your attention to something that is important - a helpful hint, or an exception to something just mentioned. The "◁" symbol is used to mark the end of a Note.   ◁

3.2 Simple Structures

In this section, we show how variables can be assigned and simple data strings may be generated. We begin, however, by explaining how S-PLUS handles simple arithmetic operations.

3.2.1 Arithmetic Operators

Arithmetic operators in S-PLUS function in the same fashion as in most programming languages and like most calculators. For example, the plus sign (+) is used for addition, the minus sign (−) for subtraction, the asterisk (∗) for multiplication, and the slash (/) for division. To raise values to a certain power, use the caret (^) or alternatively, two asterisks (∗∗). For simple expressions, multiplication and division are evaluated before addition and subtraction. However, raising to a power takes precedence over all of them.

You can experiment with the above symbols and rules by using S-PLUS as a big, expensive calculator. Suppose you want to multiply 7 by 3. Using the asterisk as mentioned above, enter

```
> 7*3
     [1] 21
```

and you are immediately shown the answer, 21.

Note The answer in the example above includes the counter, [1]. For each line of output from S-PLUS, a counter is included to number the first entry on that line of output. Thus, when variables with lots of entries are printed, it is easier to find a specific entry.   ◁

Note Now that you understand the appearance of the counter in the S-PLUS output, it will be omitted from (most of) the rest of the examples in the book for ease of presentation.   ◁

Practicing a little more with arithmetic operators, you see that you must pay attention to the order of operators. Thus, the example

```
> 7+2*3
        13
```

yields an answer of 13 because multiplication is evaluated before addition. If you really wanted the 2 to be added to the 7 before multiplying by 3, parentheses must be included to demarcate which will be evaluated first. Thus, in the example

```
> (7+2)*3
        27
```

the parentheses separate the operation that has to be evaluated before multiplying by 3. We present additional simple examples below to illustrate the rules above.

```
> 12/2+4
        10
> 12/(2+4)
        2
> 3^2                            # power operator
        9
> 2*3^2
        18
```

These examples already show how S-PLUS works. That is, commands are entered and the response follows immediately. You do not have to write a program before you start (but you may).

3.2.2 Assignments

For most applications, you will need something a little more sophisticated than a calculator. You will need to store values into variables. The action of storing a value into a variable is called assignment. In S-PLUS, the assignment of the value 2 to the variable x is written as x <- 2. Thus, the variable x is created as

```
> x <- 2
```

Normally you also need to see the value stored in a variable. This is done in one of two ways; either enter the name of the variable, or explicitly print the variable with the print function.

```
> x
        2
> print (x)
        2
```

The first method is easier and quicker, but the latter may have to be used in certain programming situations.

Note | All assignments in S-PLUS remain until explicitly removed or over-written. The rm command may be used to explicitly remove a variable. For example, rm (x) removes the variable, x.   ◁

Note | The value of a variable may be changed at any time.   ◁

As an example of these concepts, we change the value of x and then remove it with:

```
> x <- 4                          # overwrite the previous value
> x
        4
> rm (x)                          # remove x
> x
        Error: Object "x" not found
```

Note | All assigned variables are written to your hard disk in the current working directory as specified when you started your S-PLUS session. This means that you can quit S-PLUS at any time without having to remember to save your data, as it is done automatically. When you restart S-PLUS, all data from previous sessions are still intact and ready to be used. How about trying it out now?   ◁

The previous Note brings up the point that, after a while, you may begin to forget the variable names you have already used. In this case, it is very easy to use the objects function to get a list of all the variable names currently being used in the Data directory. If you used the command at this point, you would find the following.

```
> x <- 3
> objects ()
        "x"
```

The function has a few optional parameters whereby you can search a different database (and look at all the S-PLUS functions, for example) or search for only a subset of variable names. It would be good practice with the help system for you to find out about these options on your own.

If you prefer, you can think of an assignment statement in the reverse direction. Instead of x getting the value of 2 as before, you can put the value of 2 into the variable x by

```
> 2 -> x
```

Note The reverse assignment can only be made with the dash (-) and greater than sign (>), i.e. the underscore does not work in this case.   ◁

3.2.3 The Concatenate Command (c)

Up to this point you have worked only with single numbers (called scalars). Although it is possible to work in this fashion, it is often advantageous to work with vectors, which are simply ordered collections of numbers. Suppose you have the numbers 1.5, 2, and 2.5, and you want to work with the square of each value. You could save each number into a variable and square each variable, or you could put all three numbers into a vector and square the vector. The latter is much quicker if you have an easy way to combine the numbers into a vector.

The concatenate command (c) can be used to create vectors of any length. You simply have to list the elements to be combined, separating each element by a comma. To perform the simple task above, we create x and square it.

```
> x <- c(1.5, 2, 2.5)
> x
        1.5 2 2.5
> x^2
        2.25 4.00 6.25
```

The c command can also be used with variables. If we forgot a number, it could be added easily.

```
> x <- c(x, 3)
> x
        1.5 2 2.5 3
```

Note The c command can also be used with character values.

```
> y <- c("This", "is", "an", "example")
> y
        "This" "is" "an" "example"
```

The c command also works with a mixture of numeric and character values, but note that all the elements in the resulting vector will be converted to characters even though x is a numeric vector.

```
> z <- c(x, "x")
> z
        "1.5" "2" "2.5" "3" "x"
```

The consequence of converting all elements to character strings is that numeric operators may no longer be applied to the vector.   ◁

3.2.4 The Sequence Command (seq)

It is often useful to work with regular sequences of numbers. This could include the numbers from 1 to 10, the years 1988 to 1997, or the numbers from 0 to 1 using an increment of 0.1 (0.0, 0.1, 0.2, etc.). All of these examples are easily done with the seq command. The general syntax of this function is seq (*lower, upper, increment*). Our three examples are then:

```
> seq (1, 10, 1)
         1 2 3 4 5 6 7 8 9 10
> seq (1988, 1997, 1)
         1988 1989 1990 1991 1992 1993 1994 1995 1996 1997
> seq (0, 1, 0.1)
         0.0 0.1 0.2 0.3 0.4 0.5 0.6 0.7 0.8 0.9 1.0
```

It is helpful to know that the default increment used by the seq function is 1.0. Hence, we can shorten the first two examples slightly by

```
> seq (1, 10)
         1 2 3 4 5 6 7 8 9 10
> seq (1988, 1997)
         1988 1989 1990 1991 1992 1993 1994 1995 1996 1997
```

It is also helpful to note that the lower default value used by the seq function is also 1.0, so that the first example can be further simplified to

```
> a <- seq (10)
> a
         1 2 3 4 5 6 7 8 9 10
```

The best shorthand for the seq function is to use the colon (:). Simply putting the lower value, a colon, and the upper value is equivalent to the seq function with an increment of 1.

```
> b <- 1:10
> b
         1 2 3 4 5 6 7 8 9 10
```

When using the colon operator, the lack of an increment is not a problem if you are clever.

```
> 0:10 / 10
         0.0 0.1 0.2 0.3 0.4 0.5 0.6 0.7 0.8 0.9 1.0
```

Compare the output obtained here to that from the third example above.

Note Sequences can be constructed in descending order.

```
> seq (5, 1, -1)
        5 4 3 2 1
> 5:1
        5 4 3 2 1
```

◁

3.2.5 The Replicate Command (rep)

It will also be useful to generate data that follow a regular pattern. The replicate function takes a pattern and replicates it. The general form of the command is **rep** (*pattern, number of times*).

```
> rep (1, 5)
        1 1 1 1 1
> rep (c(0, 6), 3)
        0 6 0 6 0 6
> rep (c(0, "x"), 3)
        "0" "x" "0" "x" "0" "x"
> rep (1:3, 4)
        1 2 3 1 2 3 1 2 3 1 2 3
```

The second option for the number of times to replicate the pattern need not be a scalar. If it is also a vector, it should have the same number of elements as the pattern and then each element is replicated the corresponding number of times.

```
> rep (1:3, 1:3)
        1 2 2 3 3 3
```

Another interesting example, left to the reader to try and interpret, can be constructed by invoking the **rep** function twice.

```
> rep (1:3, rep (4, 3))
```

Note Sometimes the length of the desired vector is known rather than the number of times a pattern should be replicated. In this case, the **length** option can be used in place of the number of times option.

Suppose we want a vector with 10 elements using the pattern 1, 3, 2. To do this we would use the **length** option as in the following example.

```
> rep (c(1, 3, 2), length=10)
        1 3 2 1 3 2 1 3 2 1
```

◁

3.3 Mathematical Operations

We have already seen how S-PLUS can be used as an expensive calculator by simply entering the numbers and the desired operator(s). The example shown was to multiply 7 by 3 as in,

```
> 7 * 3
        21
```

In Section 3.2.2 we learned how to save numbers of interest into variables for possible future use. The above example can be redone with this idea in mind.

```
> a <- 7
> b <- 3
> c <- a * b
> c
        21
```

Vectors were introduced in Sections 3.2.3–3.2.5, along with simple ways of generating them. The question is now, what happens if we use arithmetic operators on vectors? You should try this on your own before looking at the solution in Table 3.1 below. Create any two vectors of the same length and try adding them. (We created a and b using a <- 5*(0:3) and b <- 1:4.) Is the result what you expected? (What did you expect?) In any case, Table 3.1 gives an example of using arithmetic operators on vectors.

Table 3.1. Arithmetic operations on vectors

a	b	a + b	a − b	a * b	a/b	a**b
0	1	1	-1	0	0	0
5	2	7	3	10	2.5	25
10	3	13	7	30	3.33	1000
15	4	19	11	60	3.75	50625

Note Arithmetic operators work elementwise on vectors. That is, when vectors a and b are added, the resulting vector has as its first element the sum of the first element of a and the first element of b, as its second element the sum of the second element of a and the second element of b, and so on. ◁

Note Arithmetic operators can be applied to two vectors x and y even if they have different lengths (contain different numbers of entries). However, S-PLUS issues warning messages in this case. (Do not use this unless you know what you are doing.) ◁

| Note | Arithmetic operators can be applied to a vector and a scalar.   ◁

As an example of the above Note, consider the following example.

```
> a <- seq (0, 20, 5)
> a
      0 5 10 15 20
> 2 * a
      0 10 20 30 40
```

| Note | The rules above apply to more than two vectors in exactly the same manner.   ◁

Take a and b to be defined as in Table 3.1, then see what happens with a third vector in the following example.

```
> c <- rep (2, 4)
> a + b + c
      3 9 15 21
```

If you have been trying the examples on your own as you are reading this book, you will have noticed some warning messages after the previous example. The messages are related to the fact that you have chosen (this time on purpose) a variable name that corresponds exactly to an internal S-PLUS command, namely the c function.

| Note | You should avoid using S-PLUS function names as variable names because this will generate confusion and warning messages. If you find that you have inadvertantly used an S-PLUS function name, simply remove that variable (this will not delete the function) and choose a different variable name.   ◁

We should rectify the situation and continue with the following commands. This is especially important with the variable name, c, because S-PLUS uses that command internally and warning messages will constantly appear if the variable is not removed.

```
> rm (c)
> d <- rep (2, 4)
> a + b + d
      3 9 15 21
```

If you choose not to remove the variable c above, S-PLUS will continue to use it correctly in calculations, but will endlessly annoy you with warning messages.

Everything so far, although explanatory, has not been mathematically challenging. What if we had to evaluate the function

$$f(x, y) = \sqrt{\frac{3x^2 + 2y}{(x + y)(x - y)}},$$

for many values of x and y? You could take each pair of x and y and calculate f(x,y) for each pair, but that is rather tedious. S-PLUS works in terms of vectors, and it is to your great advantage to do so as well. To approach the above problem, we need to build vectors x and y so that they contain our values of interest.

```
> x <- seq (2, 10, 2)
> y <- 1:5
> x
     2 4 6 8 10
> y
     1 2 3 4 5
```

We now define the function, f, in terms of the vectors x and y. This will have the effect of evaluating f at each combination of x and y. Thus the vector z - which we create - will have the same number of elements as both x and y.

```
> z <- ((3*x^2 + 2*y)/((x + y)*(x - y)))^(0.5)
> z
     2.160247 2.081666 2.054805 2.041241 2.033060
```

3.4 Use of Brackets

There are three types of brackets in S-PLUS: round (), square [], and curly {}. More conventionally, these are referred to as parentheses, brackets, and braces, respectively. We have already seen two uses of parentheses (for function calls and for grouping in arithmetic expressions). We will investigate the use of square brackets here, and braces are covered in Chapter 8 on programming. The different types of brackets used in S-PLUS are summarized in Table 3.2.

Table 3.2. Different brackets in S-PLUS

bracket	function
( )	For function calls like in f(x), and to set priorities like in $3 * (4 + 2)$
[ ]	Index brackets like in x[3]
{ }	Block delimiter for grouping sequences of commands, analogous to C and the begin/end in Pascal. Example: if { ... } else { ... }

Square brackets are used to index a particular vector or matrix entry. Suppose we have a vector with three elements and we want to look at or use just the first element. We simply put the number of the element of interest within square brackets, as in the following examples.

```
> x <- seq (0, 20, 10)
> x
      0 10 20
> x [1]                         # First element of x
      0
> x [2]                         # Second element of x
      10
> x [3]                         # Third element of x
      20
> x [4]                         # Doesn't exist
      NA                        # Indicates missing
```

The index reference that is enclosed within the square brackets can be a vector itself, indicating that more than one value is desired.

```
> x [1:2]
      0 10
> x [c(1, 3)]
      0 20
```

This type of notation may also be used to exclude values. Suppose we want all values except the first. This is done in the same way as above, except that the minus sign is placed before the index.

```
> x [-1]
      10 20
> x [-c(1:2)]
      20
> y <- 1:2
> x [-y]
      20
```

3.5 Logical Values

There are two logical values (also called Boolean values), TRUE and FALSE, which are used extensively for logic. On computers, these variable types are used a great deal for making comparisons.

To appreciate the usefulness of logical variables, consider the following examples. Does 3 equal 4?

```
> 3 == 4
      F
```

Obviously 3 does not equal 4, which gives us an answer of F=FALSE.

Notice that the comparison was made with two equal signs. This and other comparison symbols are summarized in Table 3.3.

Table 3.3. Symbols for logical operators

Symbol	Function
<	Less than
>	Greater than
<=	Less than or equal to
>=	Greater than or equal to
==	Equal to
!=	Not equal to

Additional examples follow.

```
> 3 < 4
        T
> 3 != 4
        T
```

The above two lines show us that 3 is less than 4 and that 3 is not equal to 4. The exclamation point (!) is used for negation, so when used in conjunction with the equal sign (as in !=), we have "is not equal to." Now consider using logical variables with vectors.

```
> x <- 1:3
> x < 3
        T T F
```

We see that 1 and 2 are less than 3 but that 3 itself is not.

Note Logical values are coded 1 for TRUE and 0 for FALSE. As with most other languages, S-PLUS stores logical variables as 1/0 values with an attached flag indicating that this variable is of type logical.

With the flag and 1/0 coding, it is possible to perform both logical comparisons and numeric computation.   ◁

```
> x <- -3:3
> x < 2
        T T T T T F F
> sum (x < 2)
        5
```

In the above example, we have used both a logical comparison (x < 2) and a numerical computation (sum (x < 2)). (Note that the sum function simply

takes the sum of the values given to it.) For the logical comparison, we see a
T=TRUE for each value (-3, -2, -1, 0, and 1) which is less than 2, and an F for
each value (2 and 3) which is greater than or equal to 2. The sum function
operates on the 1/0 coding of the T/F values. After its internal translation,
the calculation really being performed is sum $(1 + 1 + 1 + 1 + 1 + 0 + 0)$
$= 5$. We have simply counted how many elements of x were less than 2.

Extracting elements by using logical values

One of the strengths of S-PLUS is the ability to extract subsets of variables
with ease. Suppose we have a set of heights (in inches)

```
> height <- 60:68
```

and a set of corresponding weights (in pounds)

```
> weight <- c(seq (120, 155, 5), 135)
```

We might be interested in viewing a list of weights less than 140. If we use
logical values, we can see which weights are less than 140 by

```
> weight < 140
      T T T T F F F F T
```

The problem now is that we are not looking at the values themselves, but at a
T/F string representing which weights are less than 140. However, recall that
by using brackets we can specify which elements of a vector to include. One
way of specifying the elements is to use a logical vector of the same length.
In this way, all elements corresponding to a T will be included, and those
corresponding to an F will not. To solve our problem, we take the logical
vector above and supply it as the index to weight.

```
> weight [weight < 140]
      120 125 130 135 135
```

A convenient way to read this notation is "weight where weight is less than
140." Other examples of this type follow.

```
> height [height > 65]
      66 67 68
> height [height < 60]
      numeric(0)                      # No heights < 60
> weight [weight < 140 & weight != 120]
      125 130 135 135
```

The ampersand (&) is used above for the logical "and" to print weights below
140 that are not equal to 120.

The logical expression need not be restricted to the variable of main interest. Often you will be interested in a set of values of variable a based on the
values of variable b. With the height/weight example, we may be interested
in looking at the weights of all individuals whose height is over 65. Rewritten

in words, we want weight where height is greater than 65, which fits exactly into our framework of

```
> weight [height > 65]
    155 160 135
```

Or conversely, we can look at the heights of people whose weight is less than 130.

```
> height [weight < 130]
    60 61
```

Note Be careful of creating a subset of a variable when a negative value is involved. Suppose you want to look at all values of x that are less than -1. The first guess would be to use

```
> x [x<-1]
```

but the result here would be that x now contains the single value of 1. Unintentionally, the assignment operator (<-) has been used. There are two solutions to this problem: put a space between the < and the -, or enclose the -1 in parentheses, as in (-1).   ◁

3.6 Review

This review is intended to cement the ideas presented so far. If you feel the ideas up to this point are elementary and you would be bored with a review, by all means skip the review and move right to the exercises. If you have any doubt about which to do, then you would profit from going through the review. It has been structured as a detailed example using and explaining the ideas from this chapter.

We begin with some examples of using the seq and rep functions, both of which are quite useful for creating indexing and indicator variables.

```
> seq (1, 6, 1)              # 1 to 6, increment of 1
    1 2 3 4 5 6
> seq (1, 6, 2)              # 1 to 6, increment of 2
    1 3 5
> rep (1, 6)                 # Repeat 1, 6 times
    1 1 1 1 1 1
> rep (6, 1)                 # Repeat 6, 1 time
    6
> rep (1:3, 2)               # Repeat (1:3), twice
    1 2 3 1 2 3
```

We now show more examples of the principles from our first session using the Geyser data set of S-PLUS. The Geyser data set comes with S-PLUS as an example and contains data collected about the Old Faithful Geyser in the Yellowstone National Park. There are two variables: the waiting times (in minutes) and the corresponding duration (also in minutes) of the eruption. To make computations easier, we first copy the waiting and duration times into simple vectors.

```
> waiting <- geyser$waiting
> duration <- geyser$duration
```

We are now ready to do some work and start by creating a vector with the first ten waiting times.

```
> wait.first10 <- waiting[1:10]
> wait.first10
      80 71 57 80 75 77 60 86 77 56
```

We follow that by creating another vector with the next five waiting times.

```
> wait.next5 <- waiting[11:15]
> wait.next5
      81 50 89 54 90
```

Concatenate the vector with the first ten waiting times with the vector containing the next five waiting times to create a new vector with the first fifteen waiting times. It might not be the most exciting example, but it is a nice example of vector concatenation.

```
> wait.first15 <- c(wait.first10, wait.next5)
> wait.first15
      80 71 57 80 75 77 60 86 77 56 81 50 89 54 90
```

To get some practice with data editing, insert the value, 100, at the beginning of the wait.first10 vector.

```
> c(100, wait.first10)
      100 80 71 57 80 75 77 60 86 77 56
```

Do the same type of thing by putting the value, 100, between the vectors wait.first10 and wait.next5

```
> c(wait.first10, 100, wait.next5)
      80 71 57 80 75 77 60 86 77 56 100 81 50 89 54 90
```

For a little practice with subsetting, we will create vectors containing the waiting times less than fifty and also the corresponding durations.

```
> short.wait <- waiting [waiting<50]
> short.wait.duration <- duration [waiting<50]
```

The length command is used to calculate how many elements a vector (or other construct) contains.

```
> length (short.wait)
      16
> length (short.wait.duration)
      16
```

The **mean** is used to calculate the mean, or average, of the elements supplied to it. Here we calculate the mean values of several of the vectors we just created. Note that the mean duration of the short waiting times is larger than the mean for all durations.

```
> mean (short.wait)
      47.8125
> mean (waiting)
      72.31438
> mean (short.wait.duration)
      4.432292
> mean (duration)
      3.460814
> short.wait.duration
   4.600000   4.466667   4.333333   4.000000   4.600000
   4.916667   4.583333   4.000000   4.000000   4.466667
   4.450000   4.416667   4.983333   4.000000   4.333333
   4.766667
```

Logical operators can be used to create subsets as we saw above, but remember that a double equal sign (==) is used for the logical "equal to."

```
> short.wait.duration==4
        F F F T F F F T T F F F F T F F
```

Replacing one value with another is something that you most often do. Such replacements can easily be done in S-PLUS by working on the lefthand side of the assignment statement (<-). All we have to do is set up the subset of elements we want to change (adj.short.wait.dur equal to 4) and use it as the indices to the vector. We replace all durations of 4 with 4.01.

```
> adj.short.wait.dur <- short.wait.duration
> adj.short.wait.dur [adj.short.wait.dur==4] <- 4.01
> adj.short.wait.dur
   4.600000   4.466667   4.333333   4.010000   4.600000
   4.916667   4.583333   4.010000   4.010000   4.466667
   4.450000   4.416667   4.983333   4.010000   4.333333
   4.766667
```

We have seen here how we can use just a few simple commands and ideas to calculate some rather sophisticated results.

3.7 Exercises

Exercise 3.1

Calculate the first 50 powers of 2, i.e. 2, 2*2, 2*2*2, etc.
Calculate the squares of the integer numbers from 1 to 50.
Which pairs are equal, i.e. which integer numbers fulfill the
condition $2^n = n^2$?
How many pairs are there?
(Use S-PLUS to solve all these questions!)

Exercise 3.2

Calculate the sine, cosine, and the tangent for numbers ranging from 0 to
$2*\pi$ (with distance 0.1 between them).
Remember that $\tan(x) = \sin(x)/\cos(x)$. Now calculate the difference between
$\tan(x)$ and $\sin(x)/\cos(x)$ for the values above. Which values are exactly equal?
What is the maximum difference? What is the cause of the differences?

Exercise 3.3

Use the S-PLUS help routines (not the manuals) to find out how to use the
functions floor, trunc, round, and ceiling, and what they do. Predict what
each of these functions will give as an answer for the numbers -3.7 and +3.8.
Use S-PLUS to test your predictions.

3.8 Solutions

Solution to Exercise 3.1

We first create a vector (`ints`) containing the integers from 1 to 50.

```
> ints <- 1:50
```

We then create our vectors of interest by raising 2 to the power of our integers and by raising our integers to the power 2.

```
> x <- 2^ints
> x
    2 4 8 16 32 64 128 256 512 1024 2048 ...
> y <- ints^2
> y
    1 4 9 16 25 36 49 64 81 100 121 ...
```

We create a T/F vector (being the same length as both x and y) which contains a T when elements of x and y are equal and an F when they are not equal.

```
> equal <- (x == y)
> equal
    F T F T F ... F
```

Now we print the values that are common in both x and y.

```
> x [equal]
    4 16
```

We can also see which entries were the ones in common. (The second and the fourth entries of x were equal to the second and fourth entries of y.)

```
> ints [equal]
    2 4
```

The `length` function is used to determine how many elements are contained in a vector.

```
> length (x[equal])
    2
```

We can do the same calculation using the `sum` function.

```
> sum (equal)
    2
```

Solution to Exercise 3.2

It helps to notice that π is a built-in constant in S-PLUS, denoted by **pi**. If you didn't notice that, you could have simply used the approximate value of 3.14. You need to generate the set of numbers using the **seq** function and then calculate the difference between the tan and sin/cos results. Note: The values below may differ on different machines, as this is hardware dependent.

```
> x <- seq (0, 2*pi, by=0.1)
> a <- sin (x)
> b <- cos (x)
> c <- tan (x)
> diff <- c-a/b
```

The command below tells you which of the elements are truly equal to zero. Notice that many of them are not zero, despite the fact that theoretically they should be.

```
> diff == 0
   T F F T T F F T F F F F F T F T T T T
   T T F T T T F T T T T T T T T F T T T F
   T T T F F T T F T T T T T T T T F T T F T T T
```

To try to investigate this strange result, we look at the following: the total number of values we have, the number of values that are exactly equal to zero, and the largest of the differences (the absolute value of it).

```
> length (diff)
    63
> sum (diff==0)
    43
> max (abs (diff))
    1.421085e-014
```

There were a total of 20 of the 63 elements that were not zero, but the largest was only about 1×10^{-14}. For most practical calculations, an error of this magnitude is negligible. However, you should be aware of it in case such precision is important. The magnitude of such errors may vary from machine to machine.

Solution to Exercise 3.3

You have to be a little careful with the different functions used for truncating, especially if you have both positive and negative numbers. Here is a little summary of what the functions do.

floor   Rounds to the next lower integer
trunc   Cuts off all digits after the comma
round   Rounds to the nearest integer
ceiling   Rounds to the next higher integer

You should have found the solutions listed in Table 3.4. Hopefully they also match your predictions.

Table 3.4. Truncation functions

	x = -3.7	x = 3.8
floor (x)	-4	3
ceiling (x)	-3	4
trunc (x)	-3	3
round (x)	-4	4

The floor and ceiling functions are a little confusing if you have negative numbers because the next lower and next higher integers are the reverse of what you expect. The next lower integer from -3.7 is -4 and the next higher integer from -3.7 is -3.

4. A Second Session

Now that you have learned some of the most basic functions available to you in S-PLUS, it is time to move on to more advanced data structures that will allow you to complete complicated tasks easily. We begin with matrices and then branch out into more specialized structures: subsetting by index, and missing values. We close with a few new applications and a review of the material covered in the chapter.

4.1 Constructing and Manipulating Data

Matrices can be viewed as two-dimensional vectors and therefore, they have two indices instead of one. Matrices are displayed in rectangular form: the first index denotes the row, the second the column of the specific elements.

Arrays are a more general data type in S-PLUS. An array is a general structure of more or less any dimension (actually up to eight). Specific names are given to arrays with a particular number of dimensions: a zero-dimensional array is called a scalar, single value, or point, a one-dimensional array is called a vector (S-PLUS uses column vectors), and a two-dimensional array is called a matrix.

Data frames are very similar to matrices except that they allow the columns to contain different types of data, whereas a matrix is restricted to one type of data only. Lists are the most general of all the structures because they do not have to be rectangular in layout and can contain any type of data.

We consider first the matrix type, as this is one of the most important types in mathematics and statistics.

4.1.1 Matrices

A matrix can be thought of as a rectangular layout of cells in which each cell contains a value. This arrangement of information is similar to a spreadsheet, in which a cell's contents is accessed by referencing its corresponding row and column numbers. The concept of a matrix is much easier to visualize than to describe, so we will explore some simple examples.

To create a matrix, the S-PLUS function `matrix` is useful. The general syntax is as follows.

```
> matrix (data, nrow, ncol, byrow=F)
```

Note that there are four arguments. Only the first is needed. The others are optional. If neither `nrow` (the number of rows) nor `ncol` (the number of columns) is specified, as in

```
> matrix (1:6)
     1 2 3 4 5 6
```

the returned argument is a one-dimensional matrix (column vector). If only one argument of `nrow` or `ncol` is given, as in

```
> matrix (1:6, nrow=2)
     1 3 5
     2 4 6
```

the other value is determined (by simple division) and set. If both dimensions are supplied, the matrix is simply filled up. In all these situations, a dimension that does not fit the data can be specified. This is usually not what you want to do, but a matrix will be created nonetheless, and a warning will be issued.

Let's look at how this simple matrix was created and some of its characteristics. We wanted to put the integers 1 to 6 into the matrix using 2 rows. With 6 values and 2 rows, S-PLUS does the simple division $(6/2=3)$ to calculate that the resulting matrix will have 3 columns. Recall that S-PLUS works in terms of column vectors, hence it fills the first column first, then the second, and finally the third.

In addition, the matrix can be filled row-wise instead of column-wise by setting the `byrow` argument to T.

```
> matrix (1:6, nrow=2, byrow=T)
     1 2 3
     4 5 6
```

You will find that this will be necessary to do when reading in your own data. Try to guess what the following commands produce, then try it out.

```
> x <- 3:8
> matrix (x, 3, 2)
> matrix (x, ncol=2)
> matrix (x, ncol=3)
> matrix (x, ncol=3, byrow=T)
```

| Note | Matrices consist of elements of a single type. To store columns of different types in a matrix form, see Section 4.1.3 on data frames.   ◁

To help you further with matrices and their uses, let's build a more instructive example matrix. Suppose we have data on the gross domestic product (GDP, in billions of US$), population (in millions), and inflation (consumer price inflation from 1995) for several European countries.[1]

```
> data <- c(197, 8, 1.8, 1355, 58, 1.7, 2075, 81, 1.8)
> country.data <- matrix (data, nrow=3, byrow=T)
> country.data
       197    8   2.2
      1355   58   1.7
      2075   81   1.8
```

Our matrix above is not terribly useful to us in this state because we don't know which column corresponds to which variable, or which row pertains to which country. To make our lives easier, we can give names to the rows and/or columns by using the dimnames function. This is a useful way to remember what is stored in which fields. Dimnames are used to obtain a better display, and can also be used to access field entries.

Since a matrix has two dimensions, let's say n rows and p columns, the 2 vectors containing the dimnames that need to be attached to the matrix should have n and p elements, respectively. Each element is a character string describing the corresponding contents.

Example 4.1. Attaching dimnames to a matrix
In the following code, we assign row and column headings, using dimnames, to a matrix.

```
> dimnames (country.data)
       NULL                          # NULL means empty
> dim (country.data)
       3 3
> countries <- c("Austria", "France", "Germany")
> variables <- c("GDP", "Pop", "Inflation")
> dimnames (country.data) <- list (countries, NULL)
> country.data
       Austria   197    8   2.2
        France  1355   58   1.7
       Germany  2075   81   1.8
> dimnames (country.data) <- list (NULL, variables)
> country.data
        GDP   Pop   Inflation
        197     8         2.2
       1355    58         1.7
       2075    81         1.8
> dimnames (country.data) <- list (countries, variables)
```

---

[1] The data are taken from *The Economist Pocket World in Figures*, 1997 Edition.

```
> country.data
               GDP   Pop   Inflation
   Austria     197    8        2.2
   France     1355   58        1.7
   Germany    2075   81        1.8
> dimnames (country.data)
  [[1]]:
     "Austria" "France" "Germany"
  [[2]]:
     "GDP" "Pop" "Inflation"
```

Notice that with **dimnames**, the actual row and column names are entered using a list structure (covered later in this chapter). However, the important thing is that, for a matrix, there are two parts to the **dimnames** specification: the row names and then the column names separated by a comma. The value NULL is used either to remove the dimnames of the rows or columns, or to simply specify that none will be defined.

Accessing Elements of Matrices. To access a piece of information in a matrix, you need to know the corresponding coordinates from the rectangular layout. In other words, we need to know the row and column numbers of the cell of interest. We know that Austria is in the first row and that population is in the second column. To find the population of Austria, we specify the following code.

```
> country.data[1, 2]
     8
```

We discover that the population of Austria is 8 million.

> | **Note** | Data are accessed from a matrix by first specifying the row number(s) and then the column number(s). ◁

To access an entire row (column), put the number of the desired row (column) and all the corresponding column (row) numbers. For France, this becomes

```
> country.data[2, 1:3]
    GDP   Pop   Inflation
   1355   58        1.7
```

A shorthand notation for this operation is to leave out the specification of all the column (row) numbers as this is implied.

```
> country.data[2, ]
    GDP   Pop   Inflation
   1355   58        1.7
```

As you can see, this alternate notation leads to the same answer.

You can also access data through the dimnames rather than the row and column numbers. The numbers need only be replaced by the appropriate names (surrounded by quotation marks). Some examples follow.

```
> country.data["Germany", "Inflation"]
      1.8
> country.data[, "Inflation"]
      Austria   France   Germany
        2.2      1.7       1.8
```

Using these few simple principles, you should be able to access any part of your matrix.

The same methods of accessing elements or sections of a matrix can also be used to assign or overwrite. Simply specify the elements you want to redefine and assign the new value(s). We could change the population of Austria to 10 million with the following command.

```
> country.data["Austria", "Pop"]
      8
> country.data["Austria", "Pop"] <- 10
> country.data["Austria", "Pop"]
      10
> country.data["Austria", "Pop"] <- 8   # Original value
```

Rules for Computations with Matrices. You might recall, with some effort, the rules for calculation with matrices. Here is a little refresher and an introduction to how this works in S-PLUS.

To add or subtract matrices, you should work with matrices of the same dimension (S-PLUS will also work with a matrix that is a multiple of another, but beware of the answer). Thus, if matrix A has 3 rows and 8 columns, it can be added to (or subtracted from) matrix B with 3 rows and 8 columns, but not to matrix C with 5 rows and 8 columns. Matrix addition and subtraction is elementwise, meaning that the (i,j)th element of the first matrix is added to (or subtracted from) the (i,j)th element of the second. S-PLUS uses the normal symbols for matrix addition and subtraction just like for scalar addition and subtraction. The following is a simple example.

```
> A <- matrix (0:5, 2, 3)
> B <- matrix (seq (0, 10, 2), 2, 3)
> A+B
        0   6   12
        3   9   15
```

In S-PLUS, there is one big exception to the rule of having matrices' dimensions match. If you have a single number (scalar), it can be added to, subtracted from, multiplied by, or divided into a matrix of any dimension. Take the matrix, A, above, and define a scalar, d, to be any number. Try adding d+A. Is the result what you expected? Now try A/d. Given what you saw with the addition, did the division give you what you expected?

Matrix multiplication is more complicated and is not elementwise. The S-PLUS syntax to multiply two matrices is as follows.

```
> A %*% B
```

The trick to matrix multiplication is that the number of columns of the first matrix have to be equal to the number of rows of the second. Hence, in the example above, if matrix A has 2 rows and 3 columns, matrix B must have 3 rows but can have any number of columns.

| Note | You can also do matrix multiplication elementwise. To do this we simply use the normal multiplication sign (*).

```
> A*B
       0    8   32
       2   18   50
```

For the resulting matrix, the first element of A is multiplied by the first element of B and so on.   ◁

In general, division is not possible with matrices. However, a common matrix computation is to take the inverse, which is possible in S-PLUS using the solve function. More advanced work with matrices may be possible with the singular value decomposition. This function is appropriately named svd in S-PLUS.

Recall that column vectors may be multiplied (or divided), even if they both come from the same matrix. From our example with GDP, we can divide GDP by the population in the following way.

```
> country.data[, "GDP"] / country.data[, "Pop"]
       Austria     France    Germany
        24.625   23.36207   25.61728
```

Mathematical operations on vectors work in an elementwise fashion, unlike matrix multiplication.

Merging Matrices. Frequently you may discover that you have forgotten a row or column (or several), and would like to add it at a later stage. To add rows or columns to matrices, simply create a new matrix containing the missing data; and then merge it to the existing matrix. To add extra rows, use the rbind function, and to add extra columns, use the cbind function.

Suppose we forgot the variable, area (x1000 square kilometers), for the countries in our example. We could add this information to our matrix in the following way.

```
> Area <- c(84, 544, 358)
> country.data <- cbind (country.data, Area)
> country.data
```

	GDP	Pop	Inflation	Area
Austria	197	8	2.2	84
France	1355	58	1.7	544
Germany	2075	81	1.8	358

Notice that if you create a (column) vector, you can add it with no special manipulation. Also, if you give the vector the appropriate name, the dimnames are automatically updated as desired.

You add a row in exactly the same manner as with column vectors, except that you use the **rbind** function instead of **cbind**. To demonstrate how this works, we will add a row of data for Switzerland.

```
> Switzerland <- c(265, 7, 1.8, 41)
> country.data <- rbind (country.data, Switzerland)
> country.data
```

	GDP	Pop	Inflation	Area
Austria	197	8	2.2	84
France	1355	58	1.7	544
Germany	2075	81	1.8	358
Switzerland	265	7	1.8	41

You can add multiple rows/columns in the same manner, but pay careful attention to the dimensions of both the existing matrix and the one to be added.

4.1.2 Arrays

In S-PLUS, an array is a data construct that can be thought of as a multi-dimensional matrix (up to eight dimensions). It is difficult to conceive of anything beyond three dimensions, but arrays may sometimes be useful to hold multiple two- or three-dimensional matrices. One nice feature of this construct is that the rules and procedures introduced for matrices hold for arrays as well.

An **array** is defined as follows:

```
> array (data, dim)
```

The **data** must be a single expression like c(1,4,11,18,12,6) and the same for the **dimensions**. A simple array is created by the following command.

```
> array (1:24, c(3, 4, 2))
```

This produces a three-dimensional array: the first dimension has three levels, the second has four levels, and the third has two levels. When printing an array, S-PLUS starts with the highest dimension and works toward the lowest dimension, printing two-dimensional matrices at each stage. The easiest explanation of how this process works is to look at an example for yourself.

Example 4.2. Handling arrays
Here is a short example to show how arrays work.

```
> x <- array (1:24, c(3, 4, 2))
> x
    , , 1
            [,1]   [,2]   [,3]   [,4]
      [1,]     1      4      7     10
      [2,]     2      5      8     11
      [3,]     3      6      9     12

    , , 2
            [,1]   [,2]   [,3]   [,4]
      [1,]    13     16     19     22
      [2,]    14     17     20     23
      [3,]    15     18     21     24
```

If we fix the second index at the value 2, we get a 3×2 matrix.

```
> x[, 2, ]
          [,1]   [,2]
    [1,]     4     16
    [2,]     5     17
    [3,]     6     18
```

Notice that S-PLUS prints one matrix for each level of the third dimension (highest dimension) of the array. The three-dimensional array is shown in two-dimensional slices. The matrix above of dimension 3×4×2 is shown in two slices, each of dimension 3×4.

Apply. The apply function offers a very elegant way of handling arrays and matrices. It works by successively applying the function of your choice to each row (first dimension), each column (second dimension), or each level of a higher dimension. The syntax is

```
> apply (data, dim, function, ...)
```

data is the name of your matrix or array, and function is the name of any S-PLUS function that you want to apply to your data. For a matrix of two dimensions, the option dim can take the integer value of 1 or 2 to refer to the rows or columns, respectively. The option, ..., can be filled in with options to be passed on to the function being specified. We can use such a function to compute the maximum value of each variable from our European Countries example.

Example 4.3. Using `apply` on arrays

Here we use the `apply` function to calculate the maximum of each column in our country data set.

```
> apply (country.data, 2, max)
     GDP   Pop   Inflation   Area
    2075    81         2.2    544
```

The `max` function simply returns the maximum value of the values supplied to it. Remember that any function can be given to `apply` so that you can use it for many purposes.

4.1.3 Data Frames

Data frames are a relatively new construct that allow you to bind data vectors of different types together, such that the data can be accessed like a matrix (similar to data sets in SAS). The syntax for creating a data frame is as follows.

```
> data.frame (data1, data2, ...)
```

Here the ... notation simply shows the fact that you can specify as many data sets as you want.

This functionality fits in nicely with the next step in our European Countries example. You may have noticed a major difference between the countries (other than language), which we will indicate with a new character variable: namely, all the countries belong to the European Union (EU) except Switzerland. We could try to use the principles from the previous section on merging matrices, but the different data types cause the following problem.

```
> EU <- c("EU", "EU", "EU", "non-EU")
> country.data1 <- cbind (country.data, EU)
> country.data1
                GDP    Pop   Inflation   Area       EU
    Austria    "197"   "8"       "2.2"   "84"     "EU"
     France   "1355"  "58"       "1.7"  "544"     "EU"
    Germany   "2075"  "81"       "1.8"  "358"     "EU"
Switzerland    "265"   "7"       "1.8"   "41" "non-EU"
> apply (country.data1, 2, max)
        Error in FUN(x): Numeric summary undefined for mode
        "character"
        Dumped
```

The matrix was created, but all entries were converted to character values and numeric calculations are no longer possible. This may be what you want to do on a rare occasion, but more often than not, you will want to preserve the numeric nature of your matrix by creating a data frame as shown in the following example.

Example 4.4. Creating data frames
Here we make a data frame out of our European Countries data set so that we can add a character variable.

```
> EU <- c("EU", "EU", "EU", "non-EU")
> country.frame <- data.frame (country.data, EU)
> country.frame
                GDP   Pop   Inflation   Area        EU
    Austria     197    8          2.2     84        EU
     France    1355   58          1.7    544        EU
    Germany    2075   81          1.8    358        EU
Switzerland     265    7          1.8     41    non-EU
```

For the most part, data frames can be used as if they were matrices. For instance, we can calculate the maximum values of all of the numeric components of our data frame.

```
> apply (country.frame[, 1:4], 2, max)
   GDP   Pop   Inflation   Area
  2075    81         2.2    544
```

| Note | You can only apply numeric computations to the numeric variables in a data frame.   ◁

The individual variables in the above example are then referenced by either of two different notations, `country.frame$GDP` or `country.frame[, "GDP"]`. These forms of referencing are rather long and tedious, especially if your typing is not up to par. However, if you know that you are going to be working with a particular data frame, there is some relief in the form of the `attach` command. If you "attach" a data frame, you can work with the simple variable names (without the `country.frame$`) until the data frame is detached with the `detach` function.

Example 4.5. Attaching and detaching a data frame
When attaching a data frame, you must be aware of variables already existing in your session that might conflict with the name of a variable from an attached data frame. We will see this problem below.

```
> attach (country.frame)
> Pop
    Austria   France   Germany   Switzerland
          8       58        81             7
> EU
    "EU"   "EU"   "EU"   "non-EU"
```

Note that there are no column labels printed for EU. This is the local version of the variable that is being printed. The local version was originally created to add to the data frame. To see the data frame version, detach, remove the local version, and attach again.

```
> detach()
> rm (EU)
> attach (country.frame)
> EU
        Austria   France   Germany   Switzerland
             EU       EU        EU        non-EU
> detach()
> EU
        Error: Object "EU" not found
```

Note | The local version of a variable always has preference rather than a version from the attach function.   ◁

In the example above, the local version of the variable EU was printed after the attach was made. We know this because no row labels were printed. After we deleted the local version and reattached, the row labels were printed as expected. We didn't have this problem with the variable Pop because it was originally created as part of a matrix and, hence, no local version existed.

The attach function is much like the "set" command in a SAS data step.

4.1.4 Lists

Lists are a construct that allow you to tie together related data that do not share the same structure. Recall that with data frames, the data could be of different types, but had to have the same structure.

Consider again the European Countries example. Suppose we want to record all the European countries in which German is an official or major language and then associate this information with the rest of our European data. First you need to know that there are six European countries in which German is an official or major language: Austria, Belgium, Germany, Liechtenstein, Luxemburg, and Switzerland. We now continue the European Countries example by showing how to create and work with lists.

Example 4.6. Creating lists
We start by creating a vector to add to our European countries data to make a list.

```
> Ger.Lang <- c("Austria", "Belgium", "Germany",
+ "Liechtenstein", "Luxemburg", "Switzerland")
```

```
> country.list <- list (country.frame, Ger.Lang)
> country.list
[[1]]:
                     GDP     Pop   Inflation    Area          EU
        Austria    "197"    "8"      "2.2"     "84"        "EU"
         France   "1355"   "58"      "1.7"    "544"        "EU"
        Germany   "2075"   "81"      "1.8"    "358"        "EU"
    Switzerland    "265"    "7"      "1.8"     "41"    "non-EU"
[[2]]:
    "Austria"          "Belgium"       "Germany"
    "Liechtenstein"    "Luxemburg"     "Switzerland"
```

We now try to access some specific information.

```
> country.list$Ger.Lang
      NULL
> country.list[[1]]["Austria", "Pop"]
      8
> country.list[[1]]["Austria", "EU"]
      EU
      Levels:
      "EU" "non-EU"
> apply (country.list[[1]][, 1:4], 2, max)
      GDP   Pop   Inflation   Area
     2075    81         2.2    544
```

The list function works by simply "pasting" together all the pieces supplied to it. It creates one item for each supplied argument, and these are sequentially referenced using double square brackets. Notice that the syntax used with data frames (country.list$Ger.Lang) does not work with lists, and a value of NULL is returned.

Individual elements in a list may be referenced by considering the double brackets as part of the name of the structure of interest. The command country.list[[1]] extracts the first list element. The referencing is then like any matrix. If you want to see the entry in the first row, second column, add [1, 2] (population of Austria). The entire reference is in the example. When this is used with a character variable, S-PLUS gives you the additional information of the different levels for that variable (only two levels here). The apply function may also be applied to list structures as shown above. (A more thorough discussion of lists and how they work is covered in Section 8.7.)

4.2 Introduction to Missing Values

What would you do in the European Countries example if you didn't know the GDP of Austria and wanted to leave this cell blank? In S-PLUS, the cell is not simply left blank, but is filled in with the letters NA, which stand for "not available." If we go back to this example, it is quite easy to change the value to a missing value.

```
> country.frame$GDP[1] <- NA
> country.frame$GDP
        Austria   France   Germany   Switzerland
             NA     1355      2075           265
```

The whole process seems quite straightforward. However, the problem is that certain mathematical functions in S-PLUS either don't work or they give unexpected values when applied to vectors containing missing values (NAs). Take, for example, the max function used several times in the European Countries example. When applied to the vector above, we don't get the value of 2075 that you might have expected.

```
> max (country.frame$GDP)
     NA
```

S-PLUS returns the value NA as the maximum of the GDP values. In a sense, it is being conservative because the value NA might arise as a product of other mathematical operations. In this case, the value of NA being returned as the maximum of your vector would warn you that something strange happened in your calculations. In other cases though (like ours), you don't care about the missing value because you put it there to represent unknown information. This problem can be circumvented by determining which values are not missing, and keeping only the valid values to perform the desired mathematical function.

```
> gdp.na <- is.na (country.frame$GDP)
> gdp.na
      T F F F
> max (country.frame$GDP[-gdp.na])
     2075
> max (country.frame$GDP[!gdp.na])
     2075
```

We have used the is.na function to determine which GDP values are missing (only the one for Austria). Then, using index subsetting to select certain elements, we have excluded the missing values first with the minus sign, and then with the exclamation mark (used for negation). (A more in-depth look at missing values will be presented in Section 6.5.)

4.3 Putting it all Together

Our review is again an extended example intended to give you more of what you can do with S-PLUS, given the knowledge you have acquired so far. For the sake of continuity, we will continue with the Geyser data set used in the previous chapter review.

In the previous review, you copied geyser$waiting and geyser$duration into the vectors waiting and duration, respectively, to save some typing. Now you have learned that you could have done exactly the same thing by merely using the attach function.

```
> attach (geyser)
```

You now have direct access to the two column vectors of waiting and duration. To get some practice with some of the matrix concepts, you will form a matrix out of these two vectors.

```
> geyser.both <- cbind (waiting, duration)
> dim (geyser.both)
      299 2                        # 299 rows and 2 columns
```

You can use the apply function to compute on either the rows or the columns (1 for the rows, and 2 for the columns). Here you calculate the mean and minimum values.

```
> apply (geyser.both, 2, mean)
      waiting   duration
      72.31438   3.460814
> apply (geyser.both, 2, min)
      waiting   duration
          43   0.8333333
```

The minimum values above might be from corresponding readings, but more than likely, the minimum duration does not come from the minimum waiting time. You can check this by looking at the minimum waiting time and its corresponding duration. First you need the index that corresponds to the minimum waiting time.

```
> min.wait <- waiting==min(waiting)
> geyser.both[min.wait, ]
      waiting   duration
          43   4.333333
```

The conclusion is that the minimum waiting time (43 minutes) does not correspond to the minimum duration (0.83 minutes). In fact, it looks like the minimum waiting time might correspond to the longest duration.

To look at this strange relationship between waiting times and durations, you will construct two small matrices, m.short.wait containing the shortest waiting times and corresponding durations, and m.long.wait containing the longest waiting times and corresponding durations. We create a T/F vector of waiting times less than 48.

```
> short.wait <- waiting < 48
```

Let us now use the T/F vector to select specific rows from the matrix.

```
> m.short.wait <- geyser.both[short.wait, ]
> dim (m.short.wait)
      4 2
> m.short.wait
          waiting   duration
               43   4.333333
               45   4.416667
               47   4.983333
               47   4.766667
> m.long.wait <- geyser.both[waiting > 95, ]
> dim (m.long.wait)
      4 2
> m.long.wait
          waiting   duration
              108   1.950000
               96   1.833333
               98   1.816667
               96   1.800000
```

You can place matrices on top of one another with the **rbind** function.

```
> join <- rbind (m.long.wait, m.short.wait)
> join
          waiting   duration
              108   1.950000
               96   1.833333
               98   1.816667
               96   1.800000
               43   4.333333
               45   4.416667
               47   4.983333
               47   4.766667
```

Notice that the items in the matrix join are not sorted. Values in a single vector may be sorted directly using the **sort** function, as in

```
> sort (join[, 1])
      43 45 47 47 96 96 98 108
```

However, if you want to sort on more than one variable, it is best to use the **order** function to get a permutation of the indices that will give the desired ordering when used. The following is a short example of this.

```
> ord <- order (join[, 1], join[, 2])
> ord
      5 6 8 7 4 2 3 1
> ordered.join <- join[ord, ]
> ordered.join
          waiting   duration
              43    4.333333
              45    4.416667
              47    4.766667
              47    4.983333
              96    1.800000
              96    1.833333
              98    1.816667
             108    1.950000
```

We specified first the waiting times and then the durations to the order
function so that our new list would be sorted first by waiting time and then,
in case of a tie, by duration. Look at the third and fourth entries, which
both have a waiting time of 47 minutes. The row with the shorter duration
comes first in the list. The variable ord contains the ways the rows should be
ordered, and the matrix ordered.join puts them in this order.

4.4 Exercises

Exercise 4.1

Work with the European Countries data used throughout most of this chapter.

- Compute the GDP/capita and add it to the data frame.
- Compute the maximum of each of the variables in the data frame.
- Which country has the largest GDP/capita?
- Assume that the GDP is not known for France. Recalculate the GDP/capita and the maximum value for each variable.
- What has now happened to the maximum of the GDP variable? Suppress the row containing the missing value and again recalculate the maximum of each variable.
- Look carefully at the maximum area. To which country does this area belong? Is this really the largest country in our small sample? What happened when we excluded the row with the missing GDP for France? Determine how to correctly adjust for the missing GDP and still get correct results for the maximum values. (Hint: Use the help facility.)

Exercise 4.2

This exercise is really a continuation of the previous one. Once you have added the variable of GDP/capita to the data frame from the European Countries example, sort the data frame according to GDP/capita. Notice that it puts the smallest GDP/capita first and the largest at the bottom. Use the help utility to change the sort so that the largest GDP/capita is first and smallest is at the bottom (reverse sort).

Exercise 4.3

Create a list of European countries whose official or major language is French. Add this vector to the list from the European Countries example. Use S-PLUS to produce a list of countries where both German and French are either an official or a major language. To do this, you will need to create a list where either German or French is spoken and find the countries that are duplicated in this list. (Hint: Use the help utility.) What are the physical characterics of these countries?

4.5 Solutions

Solution to Exercise 4.1

We begin by defining a new data frame by adding a new variable. Notice that we have used the round function to specify that one place to the right of the decimal point will be printed.

```
> country.frame1 <- data.frame (country.frame, GDPcap =
+ round (country.frame$GDP / country.frame$Pop, 1))
> country.frame1
              GDP   Pop  Inflation  Area      EU  GDPcap
    Austria   197    8        2.2    84      EU    24.6
     France  1355   58        1.7   544      EU    23.4
    Germany  2075   81        1.8   358      EU    25.6
Switzerland   265    7        1.8    41  non-EU    37.9
```

Use the apply and max functions to calculate the maximum of each of the variables (columns). Remember that the fifth column must be excluded from the calculations because it is not numeric.

```
> apply (country.frame1[, -5], 2, max)
    GDP   Pop  Inflation  Area  GDPcap
   2075    81        2.2   544    37.9
```

We set the GDP for France equal to NA and recalculate the maximum values.

```
> country.frame1[2, 1] <- NA
> apply (country.frame1[, -5], 2, max)
    GDP   Pop  Inflation  Area  GDPcap
     NA    81        2.2   544    37.9
```

Now we determine which rows contain a missing value and exclude them from the new maximum calculations.

```
> gdp.na <- is.na (country.frame1$GDP)
> apply (country.frame1[!gdp.na, -5], 2, max)
    GDP   Pop  Inflation  Area  GDPcap
   2075    81        2.2   358    37.9
```

The problem we have encountered here is that by searching for rows with missing values, we have excluded the entire row. In our example, this corresponds to all data on France. Hence, when we calculate the maximums, we find the largest area to be 358 which is that of Germany. What we really want to do is to exclude only that one missing value and not the whole row. The simple solution is found by looking at the help documentation for the max function. We find that the additional parameter, na.rm, excludes individual missing values from analyses.

```
> help (max)
```

```
> apply (country.frame1[, -5], 2, max, na.rm=T)
    GDP  Pop  Inflation  Area  GDPcap
   2075   81        2.2   544    37.9
```

Solution to Exercise 4.2

The trick here is that we want to sort our rows according to the GDP per capita. Hence, we have to specify this variable to the order function. The order given is the order in which the rows should be printed to get the desired ordering.

```
> ord.gdpcap <- order (country.frame1$GDPcap)
> ord.gdpcap
  2 1 3 4
```

By simply putting the new ordering into the row indices of the data frame, we get the data frame with our desired sorting.

```
> pr.sort <- country.frame1[ord.gdpcap, ]
> pr.sort
              GDP  Pop  Inflation  Area     EU  GDPcap
     France    NA   58        1.7   544     EU    23.4
    Austria   197    8        2.2    84     EU    24.6
    Germany  2075   81        1.8   358     EU    25.6
Switzerland   265    7        1.8    41  non-EU   37.9
```

We can't stress enough that you must make use of S-PLUS's help facility. If you look up a related command, it often contains a hyperlink to the command you are actually interested in. Here, for example, we look at the help for the order function and it has a hyperlink to the rev function. Using a little imagination, we might think that this will give us the reverse (descending) ordering. If you click on the hyperlink, you will see that rev is the reverse ordering, which works in the same way as the order function except that it orders in descending order.

```
> help (order)
> pr.rev <- country.frame1[rev (ord.gdpcap), ]
> pr.rev
              GDP  Pop  Inflation  Area     EU  GDPcap
Switzerland   265    7        1.8    41  non-EU   37.9
    Germany  2075   81        1.8   358     EU    25.6
    Austria   197    8        2.2    84     EU    24.6
     France    NA   58        1.7   544     EU    23.4
```

We could also look at other orderings if we were so inclined.

Solution to Exercise 4.3

There are four European countries where French is an official or a major language. Of course you can still do the exercise without this knowledge. Using the `list` function, create the new vector and combine it with the original data frame and the vector of German-speaking countries. If you were to combine the vector of French-speaking countries with the `country.list` already created in the chapter, the numbering of the components would be rather strange but could be done.

```
> Fr.Lang <- c("Belgium", "France", "Luxemburg",
+ "Switzerland")
> country.list1 <- list (country.frame, Ger.Lang, Fr.Lang)
> country.list1
  [[1]]:
                    GDP    Pop   Inflation   Area        EU
      Austria      "197"  " 8"    "2.2"    " 84"      "EU"
       France     "1355"  "58"    "1.7"    "544"      "EU"
      Germany     "2075"  "81"    "1.8"    "358"      "EU"
  Switzerland      "265"  " 7"    "1.8"    " 41"  "non-EU"
  [[2]]:
    "Austria" "Belgium" "Germany"
    "Liechtenstein" "Luxemburg" "Switzerland"
  [[3]]:
    "Belgium" "France" "Luxemburg" "Switzerland"
```

There are several ways to find the countries where both French and German are spoken. The solution here is not unique. We proceed by putting the second and third components of the list into a new vector and using the `duplicated` function to find which country names have been duplicated. We found three countries where the two languages are spoken.

```
> Both.frame <- c(country.list1[[2]], country.list1[[3]])
> dup <- Both.frame[duplicated (Both.frame)]
> dup
    "Belgium" "Luxemburg" "Switzerland"
> country.list1[[1]][dup, ]
                    GDP  Pop  Inflation  Area      EU
                     NA   NA        NA    NA      NA
                     NA   NA        NA    NA      NA
  Switzerland       265    7       1.8    41  non-EU
```

Notice that the first two lines of our list containing the characteristics of the French and German speaking countries is filled with NAs. This is simply because we don't have any country characteristics for Belgium and Luxemburg.

5. Graphics

Graphs are one of S-PLUS's strongest capabilities and most attractive features. You can create basic graphs by using the menu interface, but you can also do much more. We will take a look at how graphs can be created using the full functionality.

This section gives an overview of how to create graphs and how to use the different parameters to modify elements and layouts of graphs. A short summary on how to use the functionality efficiently follows.

Please note that there are many options and functions available for working with graphs and graphical elements. This chapter provides details without covering every single option and variant. The most important commands with their functionality are treated, and many other possibilities are mentioned. If you are interested in more details, this chapter will enable you to quickly find and understand the relevant information in the manuals provided with S-PLUS or from the online help.

We would like to emphasize that you can use the functions discussed here together with the graphical user interface. You can easily create the basic graph with just a few mouse clicks, and you can call functions from the command line to enhance the graph.

Graph creation from the command line is especially useful if you generate graphs on the screen from within a function.

5.1 Basic Graphics Commands

Once a graphics window (a graph sheet) is opened, using the command

> `graphsheet()`[1]

we can send commands to the graph sheet. If we have a variable x and a variable y, we can plot the two variables against each other simply by entering

> `plot (x, y)`

It's as easy as that. The plot should now appear in the graph sheet. In the same way, a histogram of a variable x can be generated by entering

---

[1] For versions earlier than 4.0, see Section 5.3 on graphics devices

```
> hist (x)
```

Other functions to display data comprise `boxplot` for a boxplot display of one or more variables, `stem` for a stem and leaf display, or `summary` for a numerical summary of a variable.

Simple plotting using only the elementary plotting functions becomes insufficient very quickly. You might want to add lines, change the colors of the boxplot, have different symbols to plot the data, or change labels.

We devote the remainder of the chapter to the details you need to customize graphs to your exact needs.

5.2 How Graphs Work

As the graphics output is dependent on the device you send it to (a printer, a plotter, a file, or a screen), you need to specify the output device to S-PLUS. You do this by calling the appropriate S-PLUS function (you can also call it the "device driver").

From now on, all commands producing graphical output send their result to the *active* device. The active device, if not specified otherwise, is typically the last device opened. The following section explains the available devices and their usage in detail.

5.3 Graphics Devices

If you want to create graphs, you must first know which graphics device you are going to use. In general, the standard graphics device to use is a graph sheet. A graph sheet can be started by entering

```
> graphsheet()
```

at the command line.

Other graphics devices, which can be used to start specific devices, are `win.graph` under Windows, `motif`, `X11`, or `openlook` under UNIX systems. In general, opening a graph sheet is preferred.

For opening printer devices, the standard MS Windows printer device is started with

```
> win.printer()
```

Under UNIX, you typically create a PostScript file and send it to a PostScript printer. The command to use is

```
> postscript(filename)
```

Of course a standard way of printing graphs is to create the graph on the screen, and once it is finished, to click the [PRINT] button in the menu bar.

The graph is automatically converted into the printer format and sent off to the printer. Knowing about the printer device is useful, especially if a series of graphs is to be created and sent to the printer. Instead of creating a graph and clicking the [PRINT] button, creating the next graph, printing it, etc., all graphs can be sent to the printer directly.

We list the most common devices in Table 5.1.

Table 5.1. The most common output devices

System	Device	S-PLUS Function[1]	Description
General			
Screen window			
	Graph Sheet	graphsheet	opens a graphics window
UNIX			
Screen window			opens a graphics window
	X Windows	X11	... on X11 systems
	Motif	motif	... on Motif systems
	OpenLook	openlook	... on OpenLook systems
	Iris	iris4d	... on SG Iris systems
	VT 100	vt100	... on VT 100 Terminals
Printer	PostScript	postscript	creates a PostScript file
	LaserJet	laserjet	creates a LaserJet file
	HPGL	hpgl	HPGL file for HPGL printers and plotters
	text printer	printer	character based printers
Windows			
Screen window		win.graph	graphics window
Printers	standard printer	win.printer	sends output directly to the default printer
	PostScript	postscript	opens a PostScript file
	text printer	printer	character based printers

Note: Many of the printer functions need a file name as argument, in order to save the output to a printer file, like in postscript (file="graphics.ps").
[1] The functions are called by adding brackets to their names, like graphsheet().

After opening the graphics device, S-PLUS is ready to receive graphics commands. The device is initialized, some starting values are set, and the user can create graphics elements for that device by calling the appropriate functions. An easy example is to plot the Geyser data we examined before.

```
> plot (geyser$waiting, geyser$duration)
```

If you want to close the last device opened, enter

```
> dev.off()
```

To close all open devices, enter

```
> graphics.off()
```

| Note | Do not forget to close the device. If you quit S-PLUS, S-PLUS does this for you. Closing the devices is especially important if you are in the process of creating a graphics file or if you are sending graphics commands directly to a printer. Only if you close the device will the last page be ejected to the printer; otherwise, the eject page command will not be sent to the printer or added to the file, respectively. The reason for this is that S-PLUS is still ready to receive graphics commands for the same page.   ◁

| Note | Typically there is a menu bar, which includes the [PRINT] button. If you click on this button, S-PLUS generates a file from the graph on the screen and sends it to the system printer. Make sure that the correct printer is installed. On UNIX systems, S-PLUS keeps a file with the name ps.out.xxxx.ps in the current directory.   ◁

Options

Device parameters can also be supplied with the function call, like specifying the height and width of the plot. Among others, you can change the plot orientation (from portrait to landscape).

All the devices have several options that are set to a default value found in an S-PLUS initialization file. You can override the default values by supplying the parameter to the function. For example, if the postscript graphics should be 5 inches wide, 4 inches high, and stored in a file smallgraph.ps, you need to enter

```
> postscript ("smallgraph.ps", height=4, width=5)
```

5.3.1 Working with Multiple Graphics Devices

S-PLUS offers the possibility of working with more than one graphics device at a time. For interactive data analysis and presentations, it is very useful to have two or more graphics windows on the screen and to show interesting details by switching between them. The S-PLUS demo function shows impressive examples. Call

```
> demo()
```

to discover what multiple windowing techniques can be used for. Note that if working with multiple devices, all devices can be referenced by their name or by their number in the list of currently active devices. Table 5.2 overviews the most commonly used functions and provides a brief explanation of their functionality.

Table 5.2. Commands for graphics devices

Command	Effect
dev.list()	Lists all currently open devices
dev.cur()	Displays name and number of the currently active device
dev.next()	Returns the number and name of the next device on the list without activating it
dev.prev()	Returns the number and name of the previous device on the list without activating it
dev.set(n)	Sets the active device to the device number n. Can be used in combination: dev.set(dev.next())
dev.copy(*device*)	Copies a graph to a new device. *device* is the function to start the new device. Example: dev.copy(win.graph)
dev.copy(which=n)	Copies the current graph to device number n. Example: dev.copy(which=2)
dev.off()	Closes the current device
dev.print()	Sends the current graph to the printer device
graphics.off()	Closes all currently active devices
dev.ask(ask=T)	Prompts the user to confirm erasing the contents of the currently active graphics device before displaying the next graph

The functions offer great flexibility of having several graphics windows on the screen while analyzing a data set. They also offer a nice toolkit for creating an impressive demonstration of self-developed applications.

5.4 Plotting Data

After learning to prepare S-PLUS for receiving graphics commands, we proceed now to generating graphical output.

In the following section, we present the most important commands to produce S-PLUS graphs. After acquiring this basic knowledge, you should be able to create graphs with more sophisticated functions.

5.4.1 The plot Command

The plot command is the most elementary graphics command. It is used to create a new figure. The plot function calculates the width and height of the figure and acts appropriately.

The most elementary usage is, as shown before,

```
> plot(x, y)
```

to plot two variables x and y of equal length against each other. In this usage, the variable in the first position (x) is plotted along the horizontal axis and the one in the second position (y) is plotted along the vertical axis.

You can also specify x only and omit y. If x is a vector, S-PLUS plots an index vector against x (1..n, n being the length of x). If x is a matrix with two columns, S-PLUS will plot the first column against the second. If x is a list containing the elements x and y, these elements are used for plotting. Every single point of the graph is displayed in the type option *point*, as this is the default. (We will come back to this in the next section.)

You can change the character for the points by specifying the parameter pch. For example,

```
> plot(x, y, pch="P")
```

displays the data points with the character P.

5.4.2 Modifying the Data Display

If you want to display data points in a graph, there are many ways of doing it. You might display the data as dots, stars, or circles, simply connect them by a line, or both. The data might be displayed as height bars, or made invisible to set up only the axes and add more elements later. For this purpose, the function plot has a parameter type. Figure 5.1 shows examples of the most commonly used type options. Options for the type mode in which to plot are summarized in Table 5.3. The statements used to generate the example graph are listed for reference.

Table 5.3. Options for the parameter type in the plot command

Type Option	Plot Style
type="p"	Points (the default)
type="l"	Lines connecting the data
type="b"	Both (points and lines between points)
type="h"	Height bars (vertical)
type="o"	Overlaid points and connected lines
type="s"	Stairsteps
type="n"	Nothing

See Figure 5.1 (p. 63) for a display of the type options.

These are the commands used to generate Figure 5.1.

```
> x <- -5:5          # generate -5 -4 ... 3 4 5
> y <- x^2           # y equals x squared
```

```
> par  (mfrow=c(3, 2))           # set a multiple figure screen
> plot (x, y)                    # and create the graphs
> plot (x, y, type = "l")        # in different styles
> plot (x, y, type = "b")
> plot (x, y, type = "h")
> plot (x, y, type = "o")
> plot (x, y, type = "n")
> mtext ("Different options for the plot parameter type",
+ side=3, outer=T, line=-0.5)
```

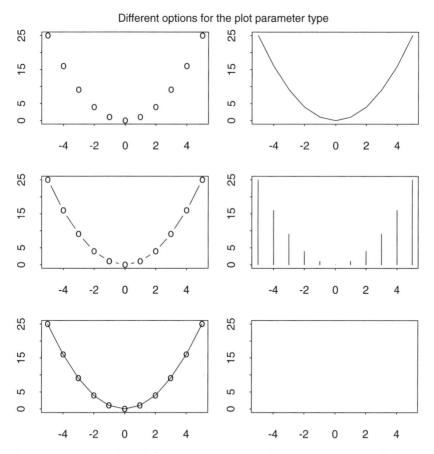

Figure 5.1. Examples of different options for the type parameter of the plot function.

5.4.3 Modifying Figure Elements

The `type` option is only one of many that can be set to modify a figure's display. You can change the labels for the axes, omit the axes, set the limits of an axis, change colors, just about anything you can imagine.

Here is an example of how several options can be combined into a single call to the `plot` function. The resulting graph is displayed in Figure 5.2.

```
> x <- seq (-5, 5, 1)
> y <- x^2
> plot (x, y, pch="X", main="Main Title", sub="Subtitle",
+ xlab="X Axis Label", ylab="Y Axis Label", xlim=c(-8, 8),
+ type="o", lty=2)
```

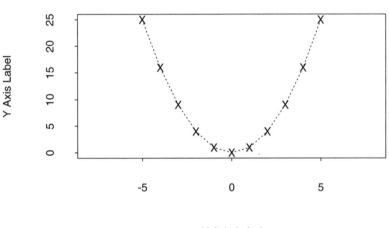

Figure 5.2. An example of a graph layout.

Figure 5.2 shows most of the main elements of a graph to give you an idea of what can be changed. Table 5.4 overviews some of the more common graphics options that may be modified.

Table 5.4. Options for plot functions

Parameter	Description
type=	Plot type, see Table 5.3
axes=T / axes=F	With/without axes
main="*Title*"	Title string
sub="*Subtitle*"	Subtitle string
xlab="*x axis label*"	x-axis label
ylab="*y axis label*"	y-axis label
xlim=c(*xmin, xmax*)	x-axis scale
ylim=c(*ymin, ymax*)	y-axis scale
pch="*"	Plot character. Example: plot (1,2,pch="?")
lwd=1	Line width. 1=default, 2=twice as thick, etc.
lty=1	Line type. 1=solid, 2=small breaks, etc.
col=1	Color, device-dependent. 0=background
box=T, box=F	Show or omit the box around the figure

Note: This is not a complete listing of the parameters available. For all details, see the online help function by entering **help (par)**.

5.5 Adding Elements to Existing Plots

This section presents some graphics *functions*, not parameters, which can be called after an initial graph is created. These functions add further elements to a graph. Figure 5.3 shows a graph with several options and functions used to modify the figure according to our needs.

5.5.1 Functions to Add Elements to Graphs

In the following example, we will examine a typical process of developing a graph. First an initial graph is created, then other elements are added in succession. S-PLUS offers many routines for adding graphics elements like axes, boxes, text, and more. Using the example from Figure 5.3 we present some important functions. Let us have a look at how Figure 5.3 was created.

Example 5.1. Creating a customized graph
We will create a customized graph as shown in Figure 5.3. To start with, we create the data to plot, then we call the basic **plot** function to set up the main graph.

```
> x <- seq (0, 2*pi, length=21)    # a sequence with 21 points
> y <- sin (x)
> plot (x,y,axes=F,box=F,type="b",pch="x",xlab="",ylab="")
```

We add axes with specified tickmarks and mixed labels and horizontal lines to the graph.

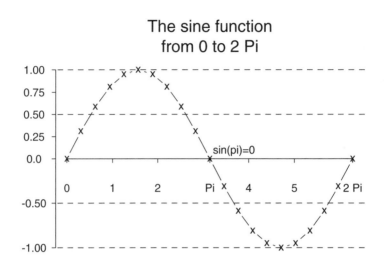

Figure 5.3. A customized graph example: a two line title, shifted x-axis, non-regular axis labels, and text added to a curve.

```
> axis (1, at=c( 0, 1, 2, pi, 4, 5, 2*pi),
+ labels=c( 0, 1, 2, "Pi", 4, 5, "2 Pi"), pos=0)
> axis (2, at=c( -1, -0.5, 0, 0.25, 0.5, 0.75, 1))
> abline(h=c( -1, -0.5, 0.5, 1), lty=3)
```

Finally, we add text to the graph in a specified position and adjusted to the left.

```
> text (pi, 0.1, " sin(pi)=0", adj=0)
> title ("The sine function\nfrom 0 to 2 Pi")
```

The commands used above (and more) are summarized in Table 5.5.

Note   Sometimes you will want to plot line segments. This is easy to do if you know that S-PLUS interrupts a line where a missing value is encountered. The line does not connect from and to the missing value. See Section 6.5 (page 123) about missing values.   ◁

5.5.2 More About abline

As abline is a very useful function for enhancing a graph with one or more lines, we give some application examples in Table 5.6, and have in fact already used it in Figure 5.3. The function abline is especially useful for drawing gridlines. If you want to have the gridlines more in the background of the picture, make them gray instead of black, such that they disturb the impression

Table 5.5. Commands that add to existing graphs

Function	Description
abline (a,b)	Add lines with intercept a and slope b to an existing figure
abline (h)	Can be used to add horizontal lines. abline(h=0:5) adds 6 horizontal lines crossing the y-axis at y=0,1,2,3,4,5
abline (v)	The analogon for vertical lines
arrows $(x1,y1,x2,y2)$	Add arrows from $(x1,y1)$ to $(x2,y2)$
axes ()	Add the x- and y-axes
axis (n)	Add an axis to a specified side. n=1 is the x-axis, n=2 the y-axis, etc.
axis $(n,\text{at}=x,\text{labels}=s,\text{pos}=y,\text{las}=m)$	Add an axis to side n as above and puts the labels s at the positions x, x and s vectors of same length. pos shifts the axis to go through the coordinate x. las=0 adds labels parallel to the axis, las=1 always adds horizontal labels, and las=2 rotates all labels by 45 degrees
box ()	Add the outer box
lines (x,y)	Add lines to an existing figure
points (x,y)	Add points to an existing figure
segments $(x1,y1,x2,y2)$	Add segments from $(x1,y1)$ to $(x2,y2)$
text $(xpos,ypos,text)$	Add *text* at the specified locations
title $("title","subtitle")$	Add a title and/or subtitle

of the whole picture less. For this purpose, you need to figure out what the color coding of the gray tones on your output device is. Another approach is to add dotted lines instead of solid ones. As usual, there are many ways to achieve a result. You need to try out different possibilities to find the most suitable one for your application.

Table 5.6. Some examples for the use of abline

Function	Description
abline (a,b)	Add a line to the current plot with axis intercept a and slope b
abline (l)	If l is a list and contains an element whose name is coefficients, uses this as arguments to abline
abline (h=x)	Draw parallel horizontal lines through all the points $(0, x)$, useful for adding a grid (if x is a vector)
abline (v=x)	Draw parallel vertical lines through all points x on the x-axis

Of course horizontal and vertical parameters can be combined. Can you think of how to create a set of parallel diagonal lines with a single S-PLUS command?

5.5.3 More on Adding Axes

Sometimes it is important to have the tickmarks sitting on the axes in the right position. S-PLUS tries to do this in a general and almost always satisfying way, but sometimes you might want to have non-equidistant tickmarks or to add a specific point on the axis. It is also possible to have the y-axis on the righthand side, and to change the labels to anything desired. These possibilities are summarized in Table 5.7.

A very common way of customizing axis layouts is to plot a graph without axes by entering, for example,

```
> plot (x, y, axes=F)
```

If, in addition, you specify box=F, the surrounding box is omitted. Now you can go ahead and add axes in the way you want. The standard routine for doing this is axis, or axes.

```
> axis (4)
```

adds an axis on side number 4, which is the righthand side (r.h.s.) of the graph (see Table 5.8 for definition of side numbers).

Note that an axis typically consists of lines. Therefore, everything you just learned about lines, like changing the thickness or the color, also applies to axes. For example,

```
> axis (4, lwd=2, col=2)
```

adds an axis on the righthand side twice as thick as the standard style, and in color number 2.

Table 5.7. Optional parameters for plotting axes

Parameter	Description
lab=c(5,5,7)	Label number of ticks on x- and y-axis and label length
las=0	Label style for axes. 0=parallel to axes, 1=horizontal, 2=perpendicular to axes
mgp=c(3,1,0)	Line of margin to place axis title, label, and axis line
xaxt="s"	x-axis type. xaxt="n" omits the x-axis
yaxt="s"	y-axis type. xaxt="n" omits the y-axis
tck=-0.02	Tick mark length. positive numbers point into the picture
xaxp=	The current tick parameters start, end, and intervals
lty=1	Line type. 1=solid, 2=small breaks, etc.
lwd=1	Line width. 1=default, 2=twice as thick, etc.

Note: The values above refer to the preset values (defaults) used if nothing else is specified.

5.5.4 Adding Text to Graphs

Adding some text to a graph is useful if specific details of a graph should be pointed out. The standard functions `text` and `mtext` do a good job. To obtain the desired result, you can supply additional parameters to these functions. A standard example for adding the text "extreme value" to a graph at the coordinates (1,2), such that the text begins on the right side of the point, is

```
> text (1, 2, "extreme value", adj=0)
```

Sometimes the text overlaps or touches the point on the graph just slightly. In such a case, add an extra space to the beginning of the text. The optional parameters of `text` and `mtext` offer more possibilities. They are listed in Table 5.8.

Table 5.8. Optional parameters for adding text

Parameter	Description
`text` and `mtext`: Text in a figure	
`adj=0.5`	Text adjustment. 0=left justified, 0.5=centered, 1=right justified
`cex=1`	Character expansion relative to standard size. cex=2 draws characters twice as big
`col=1`	Color to plot in. 0 is the background color
`crt=0`	Character rotation in degrees, counterclockwise from horizontal
`srt=0`	String rotation in degrees, counterclockwise from horizontal
`font=1`	Font for plotting characters, depending on the output device
`mtext` only: Text in the margin of the figure	
`side=`n	Side to add text to. 1=bottom, 2=l.h.s., 3=top, 4=r.h.s.
`outer=`T/F	Should the text be placed on the outer margin of the whole figure? If `outer=T`, text can be placed on top of a layout figure. For an example, see Figure 6.15 (page 129)
`line=`n	Places the text toward the margin of the figure. Negative numbers lie inside the figure

Note: Text is typically added by using the functions `text` or `mtext`. Some devices like text terminals cannot provide all the functionality like rotating text.

5.6 Setting Options

We have already seen that almost all graphics functions accept parameters for setting options like colors, character size, or string adjustment. There are many more parameters and almost all of the functions generating graphics output accept them as arguments. They can also be set generally, such that they become the system default for all succeeding graphics commmands. For doing this, we can use the **par** function.

Check the different options that can be set by **par**. Enter the command

> par()

to see all options and how they are currently set. The help pages provide a very detailed explanation.

We supply a brief overview of the most common parameters and their possibilities in Figure 5.4 and Table 5.9. If changed using the **par** function, these settings become global from then on. Nevertheless, if a new session is started they are set back to the original system defaults.

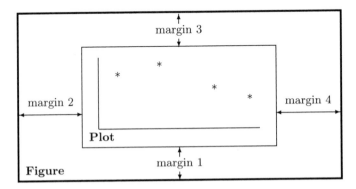

Figure 5.4. The S-PLUS definition of figure and plot regions.

Table 5.9. Layout parameters

Parameter	Description
fin=c(m,n)	Figure size in inches. m=width, n=height
pin=c(m,n)	Picture in inches, as in fin
mar=c(5,4,4,2)+0.1	All margins in lines
mai=c(1.41,1.13,1.13,0.58)	All margins in inches
oma=c(0,0,0,0)	Outer margin lines
omi=c(0,0,0,0)	Outer margin in inches
plt=c(0.11,0.94,0.18,0.86)	Coordinates of plot region as fraction of the figure region
usr	x-axis and y-axis minimum and maximum. Usually used for querying a graph's boundaries
mfrow=c(m,n)	Multiple figure layout, row-wise plotting, to generate a (m,n) matrix of pictures
mfcol=c(m,n)	Multiple figures, column-wise plotting

Note: In the left column the letters m and n are to be replaced by integer numbers. Settings given are the system default values.

The parameters referenced in Table 5.9 can be set by using the `par` function, like in the following example where the outer margin is set to be equal to 1 line on all sides, for all successive graphs.

```
> par (mar=c(1, 1, 1, 1))
```

A setting can be retrieved by entering

```
> par ("mar")
```

or

```
> par()$mar
```

If you run S-PLUS under Windows, these settings can also be changed by using the menu bars on top of the S-PLUS window.

| Note | You can add your own parameters to the system settings using an unused variable. The command

```
> options ()
```

shows all settings not related to the graphics subsystem. A new variable `just.2c.if.it.works` can be set to TRUE by entering

```
> options (just.2c.if.it.works=T)
```

at the prompt. Your own functions can then use this parameter setting by asking for its setting as follows:

```
> options ("just.2c.if.it.works")
     T
```

◁

5.7 Creating Fancy Graphs: The Most Important Commands

When you try out S-PLUS's graphics capabilities, you will discover that it creates good looking graphs most of the time. The graphical parameters we looked at in this chapter are preset with meaningful values. Nevertheless, you might want to change the default values. This section deals with the most common cases to demonstrate how to achieve satisfying results quickly.

If you use a multiple figure layout, the system may not always choose the layout well (the size of the characters, the white space between figures, etc.). We now show, for the standard layouts, how parameters are set to generate the graphics wanted.

To illustrate, a 2x2 graphics layout can be created with

```
> par (mfrow=c(2, 2))
> x <- 1:100/100:1
> plot (x)
> plot (x, type="l")
> hist (x)
> boxplot (x)
```

If you look at this graph on the screen, it might satisfy you. But if you rescale the plot to save it to a PostScript file, for example, the result is sometimes undesirable. (If you want to include a graphics file in a text document, see section 11.2 for further details.) Use the command

```
> postscript ("example.ps", height=5, width=5)
```

to create a PostScript file of width and heigth 5 inches (1 inch equals 2.54 cm) and reissue the commands above. They now create a PostScript command file. Nothing is visible on the screen or in the command window. Enter

```
> dev.off()
```

to close the last device opened. Send the file to your printer to see the result. This depends on the local installation, but for UNIX workstations, the command

```
lpr example.ps
```

often works, whereas under DOS/Windows, the command

```
print /b example.ps
```

does this job.

Note Sure enough, you need to have a PostScript printer to print out a PostScript file. Otherwise, use your appropriate device driver (see Section 5.3).   ◁

To change the size of the characters according to your needs, you must set the parameter cex. This occurs in relation to the standard character size. To reduce the character size to 60 percent of the original size, enter

```
> par (cex=0.6)
```

Very often the space between the figures is too large, as this is given in absolute values (in inches) and in lines of characters. To define the absolute distance between the single plots, the command

```
> par (mai=c(x1, x2, x3, x4))
```

helps. Here, $x1$ is the space left free for the lower side's border (the space between the overall limit of the picture and the frame of the figure itself), $x2$

refers to the left, $x3$ to the upper, and $x4$ to the right side. So to set a margin of half an inch around the figure for title, axis, and labels, use

```
> par (mai=c(1/2, 1/2, 1/2, 1/2))
```

If you now get messages like

```
no room for x-axis
no room for y-axis
```

you might want to allow the single pictures to extend their limits. The parameter **mar** specifies the maximum text lines allowed around a figure, also counterclockwise starting at the x-axis side. For example, a good setting is often

```
> par (mar=c(2, 2, 1, 1))
```

such that there is a space of two lines on the lower and the left side of the graph, usable for axis labels, and a single line on top and on the left of the graph.

Note A graph created on a Graph Sheet can be exported by using the menu function [EXPORT GRAPH] in the [FILE] menu. This option allows the specification of a variety of different formats, but not the size of the overall figure and other options. ◁

5.7.1 Figure Layouts

Instead of dividing the page into several graphs of equal size, like we just did by using `par(mfrow=...)`, we can also put several figures of different sizes onto the same page.

We need to open a new graphics page, which is done by entering the command

```
> frame()                        # open next graph
```

Note Using `frame()` in a figure layout created with `par(mfrow)` skips to the next field and all following commands affect the next field. By entering `frame()` twice, one field is left blank. ◁

After having opened a new graphics page, the coordinates of the plot are set to (0,0) for the lower left corner and (1,1) for the top right corner. Let us now plot the Geyser data set that comes with S-PLUS. If you want to know more details about it, see the S-PLUS documentation.

We want to set up a figure such that a scatterplot of the data appears as the main plot with the addition of a histogram of the x-data on top of the scatterplot and a boxplot of the y-data to the righthand side of it.

We decide that the plot itself should have the lower left corner coordinates (0,0) and the top right corner coordinates (0.7, 0.7). Therefore, the figure on top of the data plot has the lower left corner (0.0, 0.7), and the upper right corner is (0.7, 1.0). For the vertical boxplot on the right side, we get the lower left corner (0.7,0) and the upper right corner (1.0, 0.7).

The layout we have in mind looks like the one in Figure 5.5.

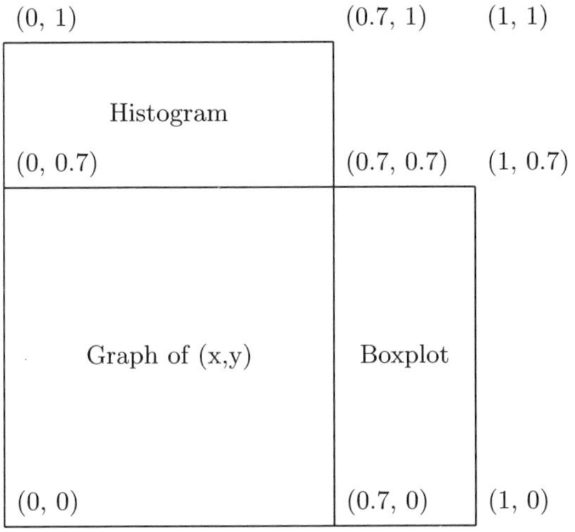

Figure 5.5. A customized graph layout.

By experimenting with the different sizes and the borders, we discover that it is better to have the figures overlapping a little bit. In order to get a smaller figure like in Figure 5.6, we adjust the size of the characters too.

Finally, we use the following commands to generate Figure 5.6.

Example 5.2. Creating a figure layout
We generate Figure 5.6 with the following commands.

```
> frame()
> par (fig=c(0.0,0.7,0.0,0.7), mar=c(4,4,2,2), cex=0.7)
> plot (geyser$waiting, geyser$duration, pch="*",
+ xlab="Waiting Time", ylab="Duration of Eruption")
> title ("\nOld Faithful Geyser Data Set", cex=0.5)
> par (fig=c(0.0,0.7,0.65,1), mar=c(2,4,2,2), cex=0.7)
> hist (geyser$waiting)
> par (fig=c(0.6,1,0.0,0.7), mar=c(4,3,2,2), cex=0.7)
> boxplot (geyser$duration)
```

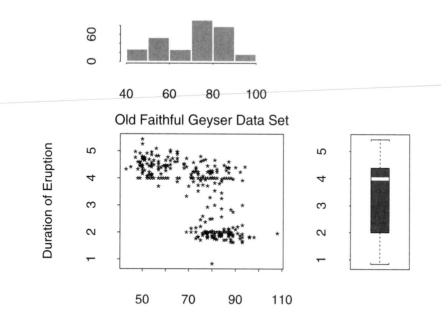

Figure 5.6. A customized display of the Geyser data set.

Summary

To summarize, here is a short listing for creating a 2x2 figure layout and storing it in a PostScript file. The parameters you typically need to modify for customizing border width, character size, etc. are used here.

```
> postscript ("example.ps", height=5, width=5)
                                   # open a postscript file
> par (mfrow=c(2, 2))              # create a 2x2 layout
> par (cex=0.6)                    # reduce character size
> par (mai=c(1/2, 1/2, 1/2, 1/2)) # change margins to 1/2 inch
> par (mar=c(2, 2, 1, 1))          # change margins to 1,2 lines
> x <- rnorm (100)                 # generate some data
> plot (x)                         # create first graph
> plot (x, type="l")               # create second graph
> hist (x, cex=0.5)                # create third graph
                                   # with different character size
> boxplot (x)                      # create fourth graph
```

> `dev.off()`   # close device (file)

If you want to restore the original graphics device status prior to your layout and other parameters, see Section 10.2.1 (page 213) on how to store and restore graphics settings.

Note You might be interested to know that you can also store S-PLUS graphs in variables. Send the command to produce the plot as an argument to the function `graphics` and redirect the output into a variable, like in the following example:

> `xygraph <- graphics (plot (rnorm (100), rnorm (100)))`

Nothing appears in the graphics window now, but the graph is stored as a graphical object in the variable `xygraph`. Enter now

> `xygraph`

or

> `print (xygraph)`

and the variable is printed, which means that the graph is displayed in the graphics window.

This functionality is currently device-dependent and not portable to devices other than the one you were producing it for, because the coordinates are stored in absolute values. Nevertheless, if it takes your system quite some time to produce a graph, you might want to store the graph in a variable so you can reaccess it immediately without any calculation.   ◁

5.8 Exercises

Exercise 5.1

In a single figure, plot the functions sin, cos, and sin+cos with different colors and line styles. Use 1000 points in the interval $[-2\pi, 2\pi]$ and label the figure with a title, subtitle, and axis labels.

Exercise 5.2

Construct two vectors, x and y, such that the S-PLUS command

```
> plot (x, y, type="l")
```

creates the following figures:
  a) a rectangle or square
  b) a circle
  c) a spiral.
Hint: Do not think of y as being a function of x (in a mathematical sense), but think of how the trace of the figure can be created.
Hint for drawing circles: you might use the property that in polar coordinates, a point $(\sin(x), \cos(x))$ lies on the unit circle with origin $(0, 0)$ and radius 1.

Exercise 5.3

We consider the so-called Lissajous figures. They are defined as

$$z(x) = \begin{pmatrix} \sin(ax) \\ \sin(bx) \end{pmatrix} = \begin{pmatrix} z_1 \\ z_2 \end{pmatrix},$$

or, in different notation,

$$z_1(x) = \sin(ax)$$
$$z_2(x) = \sin(bx),$$

a, b positive integers, x between 0 and 2π.
Plot z_1 against z_2, using lines to connect the points, and choose different pairs of values for a and b.
Plot several figures on a single sheet.
On what do the forms of the curves depend?
Compare, for example, the figures for $(a,b) = (3,4)$, $(3,6)$, and $(6,8)$.

5.9 Solutions

Solution to Exercise 5.1

What we need to create this graph is a sequence of x-values, for which we calculate the values sin(x) and cos(x). If several functions are to be plotted within one picture, the ordering of plotting the figures is important. The arguments to the plot function, usually the first curve, determine the boundaries of the figure. The following curves are then added to the existing picture within the existing limits. If you want to see more of the sin(x)+cos(x) curve than that which fits within the boundaries of sin(x), extend the y-limits by using the option ylim of the plot function or - and this is easier - plot the curve sin(x)+cos(x) first.

```
> x <- seq (-10, 10, length=1000)
> plot (x, sin(x)+cos(x), xlab="x values", ylab="y values")
> lines (x, cos(x), lty=2, col=2)
> lines (x, sin(x), lty=3, col=3)
> title ("Trigonometric Functions"
+ "sin(x), cos(x), and sin(x)+cos(x)")
```

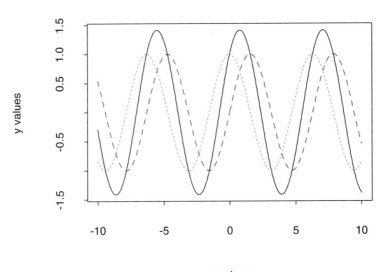

Figure 5.7. Trigonometric functions.

Solution to Exercise 5.2

The idea of this exercise is to learn how to think in terms of geometric figures and not in terms of mathematical functions, where y is a function of x.

a) A rectangle is a simple figure consisting of four lines. In order to draw a rectangle, we need to create two vectors x and y containing the coordinates of the corners. These vectors must have *five* elements, as the line has to return to the first point to complete the figure.

```
> x <- c(1, -1, -1, 1, 1)
> y <- c(1, 1, -1, -1, 1)
> plot (x, y, type="l", box=F, axes=F, xlab="", ylab="")
> title ("A rectangle")
```

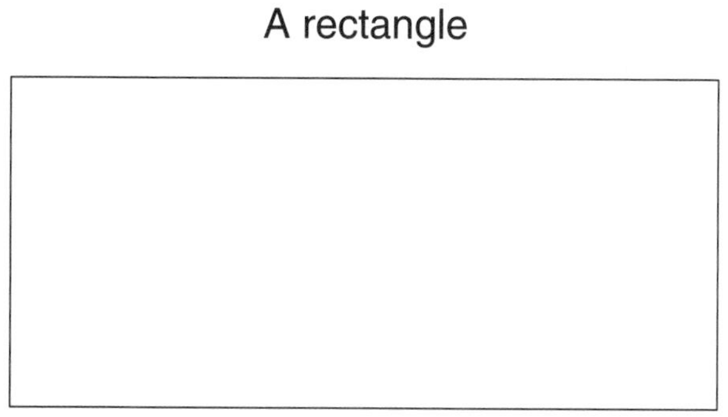

A rectangle

Figure 5.8. Geometric figures (a): A rectangle.

b) It is possible to come up with several solutions for drawing a circle. The easiest one to program is perhaps to use polar coordinate transformation. We use the fact that a point with the coordinates $(\sin(x), \cos(x))$ lies (for all x) on the unit circle with radius 1 (because $\sin^2(x) + \cos^2(x)=1$). Then we simply create a sequence from 0 to 2π and plot $\sin(x)$ against $\cos(x)$.

Another approach is to use $x^2 + y^2 = 1$ as a description for a circle. Then we have the solutions $y = \sqrt{1 - x^2}$ and $y = -\sqrt{1 - x^2}$. Try to use these formulas to draw the circle.

Finally, if you plot the vector x against y, you will discover that the resulting graph is not a circle but an ellipsoid. To obtain a circle, we need to tell S-PLUS that this plot's boundaries should be quadratic instead of rectangular. For this purpose, we set the size of the plot, the parameter `pin` (picture in inches), in order to have a quadratic picture shape.

```
> z <- seq (0, 2*pi, length=1000)
> x <- sin (z)
> y <- cos (z)
> par (pin=c(2, 2))
```

```
> plot (x, y, type="l", box=F, axes=F, xlab="", ylab="")
> title ("A circle")
```

A circle

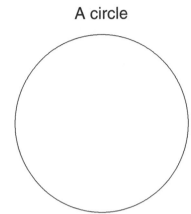

Figure 5.9. Geometric figures (b): A circle.

c) The spiral is simply made by revolving several times around the circle and changing the radius over time. The windings depend very much on the division. By changing the start, the end, or the division factor, you get very different pictures.

```
> z <- seq (6*pi, 32*pi, length=1000)
> x <- sin (z) / (0.1*z)
> y <- cos (z) / (0.1*z)
> plot (x, y, type="l", main="A spiral with 13 windings",
+ box=F, axes=F, xlab="", ylab="")
```

A spiral with 13 windings

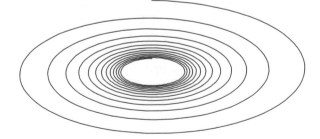

Figure 5.10. Geometric figures (c): A spiral.

Solution to Exercise 5.3

We first create the layout of the graphics window (open a window first if you haven't already). Then we create a sequence from 0 to 2∗π and store this sequence in x.

It is not necessary to define values a and b and store sin(a∗x) and sin(b∗x) in variables before plotting them, because we don't need these data any more after plotting them. Instead, we supply these expressions directly to the plot function.

```
> par (mfrow=c(2, 2))
> x <- seq (0, 2*pi, length=1000)
> plot (sin (3*x), sin (6*x), type="l")
> plot (sin (3*x), sin (8*x), type="l")
> plot (sin (3*x), sin (11*x), type="l")
> plot (sin (7*x), sin (8*x), type="l")
```

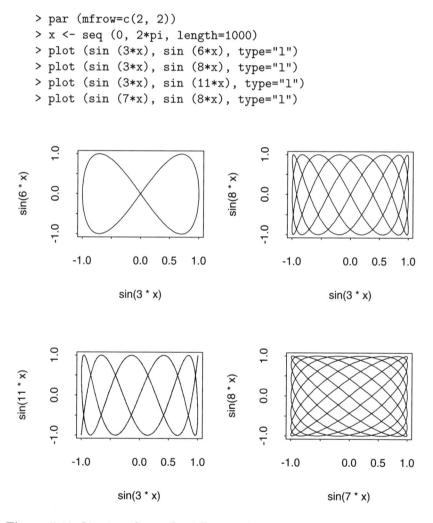

Figure 5.11. Lissajous figures for different values of a and b.

6. Exploring Data

In the preceding chapters, we have laid the foundation for understanding the concepts and ideas of the S-PLUS system. We explored basic ideas and how to use S-PLUS for performing calculations, and we have seen how data can be generated, stored, and accessed. Furthermore, we also looked at how data can be displayed graphically. All this will be useful as we explore real data sets and learn how to use the existing functionality of S-PLUS in this chapter. We will explore data sets that come with S-PLUS, specifically the Barley and Geyser data sets.

Rather than presenting a list of available statistical functions, we will go through a typical data analysis as a way of introducing the more useful and common commands and the kind of output we'll encounter. We chose to use S-PLUS data sets so you can follow along with the analysis we present and complete the exercises at the end of this chapter.

6.1 Univariate Data Exploration

In this section we will explore the different variables contained in the Barley data set. We will analyze the variables in one dimension, or in other words, univariate. The analysis of the dependence between the variables and the exploration of higher dimensional structure follows later on. We divide the data analysis into the two categories, "descriptive" and "graphical" exploration.

The Barley data set. The Barley data are measurements of yield in bushels per acre at different sites. The analysis comprises 6 sites planting 10 different varieties of barley in two successive years, 1931 and 1932. The data set therefore contains 120 measurements of barley yield. Our main goal will be to determine reasons for the different barley yields given by the different variable constellations, like the 1931 harvest of the fifth variety on site 4, and the 1932 harvest of the seventh variety at the same site.

6.1.1 Descriptive Statistics

Exploratory data analysis (EDA) is an approach to investigating data that stresses the need to know more about the structure and information inherent in the data. The methods used with this approach are referred to as descriptive, as opposed to confirmatory. Descriptive simply means that simple summaries are used to describe the data: their shapes, sizes, relationships, and so on. Examples of descriptive statistics are means, medians, standard deviations, histograms, scatterplots, ranges, and so on.

Given the basic information about the Barley data, the following analysis is intended to gain more information and structural knowledge about the numbers we have.

A typical place to begin is of course looking at the data. If the data set is small, we can easily look at it by simply printing it out. Our data set is almost too large to look at. If you type

```
> barley
```

at the S-PLUS prompt, you will get a display of 120 lines and 4 columns. We randomly pick some rows of the data matrix.

```
> barley[c(2, 17, 64, 70, 82, 98, 118), ]
```

	yield	variety	year	site
2	48.86667	Manchuria	1931	Waseca
17	29.66667	Svansota	1931	Grand Rapids
64	32.96667	Manchuria	1932	Crookston
70	26.16667	Glabron	1932	Crookston
82	32.06666	Velvet	1932	Crookston
98	44.70000	No. 462	1932	Waseca
118	35.90000	Wisconsin No. 38	1932	Crookston

This already shows us what the data look like. The yield is the number of bushels per acre, as we know from the data description (have another look at the manual or the online help pages to learn more about it). It is a decimal, and not always an integer. The number of bushels harvested was probably divided by the area of the corresponding plot of barley. The second variable, variety, contains a string describing the name of the variety of barley. The year is either 1931 or 1932, denoting the year of planting and harvesting. Finally, the site variable contains the name of the site from which the data originates.

We can determine this information in a more structured format by using the summary function, which summarizes each variable on its own.

```
> summary(barley)
        yield              variety      year              site
   Min.:14.43       Svansota:12   1932:60   Grand Rapids:20
   1st Qu.:26.87       No. 462:12   1931:60       Duluth :20
   Median:32.87     Manchuria:12             Univ Farm :20
   Mean:34.42        No. 475:12                  Morris :20
   3rd Qu.:41.40        Velvet:12               Crookston :20
   Max.:65.77       Peatland:12                 Waseca :20
                     Glabron:12
                     No. 457:12
                  Wisc No.38:12
                       Trebi:12
```

In fact, the summary function shows per default up to seven different values for factor variables, such that entering the command as above gives a slightly compressed output. To get exactly what you see above, we used summary(barley, maxsum=11). The argument maxsum is the parameter setting the maximum number of rows displayed in the summary function.

Now, we investigate the data further. To have easier access to the variables contained in barley, we attach the data set.

```
> attach (barley)
```

This gives us direct access to the variables yield, variety, year, and site.

How about having a look at our main variable yield? Let us determine what the distribution looks like by using the stem and leaf display.

```
> stem (yield)
        N = 120 Median = 32.86667
        Quartiles = 26.85, 41.46666

        Decimal point is 1 place to the right of the colon
        1 : 4
        1 : 579
        2 : 0011122223333
        2 : 5556666666677777777889999999
        3 : 000000111222222333344444
        3 : 5555667777888899
        4 : 000112223334444
        4 : 567777779999
        5 : 00
        5 : 5889
        6 : 4
        6 : 6
```

The data are ordered and categorized by a base times a power of ten. The stem and leaf display tells us that the data are put into categories 10, 20, 30, .., 60, because "the decimal point is 1 place to the right of the colon." The lefthand side displays the category (or interval), and the righthand side of the colon has one digit per observation, displayed by the first digit following the category digit. We can immediately see that the smallest values are 14, 15, 17, and 19 (rounded, of course); and by far the largest values are 64 and 66 bushels per acre. Most of the yields are in the high twenties and low thirties, and the distribution looks rather symmetric, having a longer tail toward the larger values (in other words, it is right-skewed).

The stem and leaf plot can be viewed as a sort of sideways histogram from which we can see that the data approximately exhibits the bell-shape typical of a Normal or Gaussian distribution with no extreme values or outliers. As with the histogram, the display sometimes strongly depends on the categorization layout chosen. For glimpsing the distribution and detecting extreme values in the set, the above display is sufficient. However, if we want to determine whether the distribution is approximately bell-shaped like a Normal distribution, we would need to display it with different categorizations or use more sophisticated techniques.

If we want to apply the `stem` display to the other variables of barley, we would get an error message.

```
> stem (variety)
        Error in Ops.ordered(x, 0):
        "*" not meaningful for ordered factors
        Dumped
```

What we realize here is that S-PLUS behaves somewhat intelligently. It "knows" that `variety` is a factor variable with values like "Svansota" and "Manchuria" and that a stem and leaf display for such a variable does not make much sense. Therefore, S-PLUS simply refuses to do the desired display and tells us that our attempt is not meaningful for factors; in fact, the error message comes from an operation inside the `stem` function.

If we want to investigate the skewness of the `yield` variable further, we could use the `quantile` function and calculate the quantiles at 10, 20, 30, ..., 90%.

```
> quantile (yield, seq (0.1, 0.9, by=0.1))
        10%       20%     30%       40%        50%
   22.49667     26.08   28.09   29.94667   32.86667
        60%       70%     80%       90%
   35.13333   38.97333   43.32   47.45666
```

This shows us that the 10% quantile (22.5) is about 10 units away from the median (32.9), whereas the 90% quantile is 47.5, being about 15 bushels per acre away from the median. For a symmetric distribution, the two quantiles would be about the same distance away from the median. A measure of the data spread is the interquartile distance, the difference between the 25% and the 75% quantile.

```
> quantile (yield, c(0.25, 0.75))
      25%   75%
   26.875  41.4
```

This tells us that approximately 50% of the data lie within the interval of 26.9 to 41.4 bushels per acre. Taking the median 32.87, we are not so convinced any more about the symmetry of the distribution of the yield variable.

Another measure of spread is the variance, or the standard deviation. S-PLUS does not have a built-in function for calculating the standard deviation, so you have to calculate it by taking the square root of the variance.

```
> sqrt (var (yield))
   10.33471
```

We can also use what we have learned about selecting subsets from data. We can calculate some interesting statistics by conditioning on certain values. Let us check if the barley harvest was very different in 1931 and 1932. Selecting only the values of yield that were obtained in 1931, we obtain

```
> summary (yield[year==1931])
   Min.  1st Qu.  Median   Mean  3rd Qu.   Max.
   19.7    29.09    34.2  37.08    43.85  65.77
```

and for 1932, we obtain

```
> summary (yield[year==1932])
    Min.  1st Qu.  Median   Mean  3rd Qu.   Max.
   14.43    25.48   30.98  31.76     37.8  58.17
```

As we can see, all the numbers are larger for 1931. The 1932 harvest seems to have been much worse than the 1931 yield, measured in bushels per acre.

We could determine the "most fruitful" sites for both years to see if they were the same in both. A good idea is to select the sites above the 90% quantile for both sites. We get the 90% quantile for both years as

```
> quantile (yield[year==1931], 0.9)
         90%
    49.90334
```

```
> quantile (yield[year==1932], 0.9)
       90%
     44.28
```

and retrieve all rows of the Barley data set where the yield in 1931 lies above the 90% quantile.

```
> barley[yield > 49.90334 & year==1931, ]
              yield     variety   year      site
         8  55.20000     Glabron   1931    Waseca
        20  50.23333      Velvet   1931    Waseca
        26  63.83330       Trebi   1931    Waseca
        32  58.10000     No. 457   1931    Waseca
        38  65.76670     No. 462   1931    Waseca
        56  58.80000  Wisc No.38   1931    Waseca
> barley[yield > 44.28 & year==1932, ]
              yield     variety   year      site
        86  49.23330       Trebi   1932    Waseca
        87  46.63333       Trebi   1932    Morris
        98  44.70000     No. 462   1932    Waseca
        99  47.00000     No. 462   1932    Morris
       116  58.16667  Wisc No.38   1932    Waseca
       117  47.16667  Wisc No.38   1932    Morris
```

Remember what we have learned about selecting data from matrices? In the example above, we query if yield is larger than 44.28 and year is equal to 1932, giving us a vector of TRUE/FALSE elements. These are used as row indices for the matrix barley, and the column index is omitted to get all columns. All the rows of barley are returned where it is TRUE that yield is larger than 44.28 and year is equal to 1932.

The output above reveals that Waseca seemed to have an incredible harvest in 1931 compared to the other sites. 1932 wasn't that bad for Waseca either, but Morris also had a good year. It turns out that both sites had their big yield in 1932 for the same varieties, Trebi and No. 462. This might lead to further investigation of the surprising observations we just made, but let's continue using more of the S-PLUS functionality.

A useful tool for doing analyses like the ones we just did is the by function. It applies functions to data by first splitting the data up into subcategories. We can now calculate the summaries for the Barley data by year, using the by function.

```
> by (barley, year, summary)
```

This provides a summary of the whole data set, but in the form of two summary tables, one for 1931 and one for 1932. If we want S-PLUS to display more than just seven lines (because we have ten varieties, as mentioned above), we can add the maxsum=11 setting as an additional parameter to the by function. Similar to the call to summary above, where we entered

```
> summary (barley, maxsum=11)
```

we can now enter

```
> by (barley, year, summary, maxsum=11)
```

to obtain the same formatting. In the same way, you could calculate the summary statistics or other figures for each site or each variety separately, without using a loop over the sites or varieties.

You have seen that a combination of only a handful of functions offers many ways of looking at a data set. Table 6.1 summarizes the S-PLUS functions and some of their parameters we have just seen. In addition, it contains a few more functions that might be helpful in other applications.

Table 6.1. Univariate descriptive statistics functions

S-PLUS Function	Description
quantile	Quantiles of data
mean	(Optionally trimmed) mean of data
median	Median of data
stem	Stem and leaf display
var	Variance of data or, if two variables are supplied, covariance matrix
by	Applies a function to data split by indices
summary	Summary statistics of an object
apply, lapply, sapply	Calculations on rows or columns of matrices and arrays (apply), and on components of lists (lapply, sapply)
aggregate	Aggregate data by performing a function (like mean) on subsets

Note If you want to use the by function to calculate the mean value of the yield for each year, you would encounter a little difficulty.

```
> by (yield, year, mean)
        Error in as.double:
        Cannot coerce mode list to double: .Data = list(..
        Dumped
```

The problem arises because the **mean** function is not object-oriented. In detail, by passes the **barley** elements - which are internally stored as lists - to **mean**, which in turn does not know what to do with them. Entering

```
> mean (list (123))
```

does not work either. A quick solution is to convert the list structure by using unlist before calculating the mean. We write a little function

```
> newmean <- function (x) { mean (unlist (x)) }
```

and use our function **newmean** instead.

```
> by (yield, year, newmean)
```

◁

6.1.2 Graphical Exploration

We have gained a basic impression and some insights into the Barley data set structure by studying some summary figures. The next step will lead us into graphical analyses. We will examine some displays of one variable at a time, but as most of the graphics functions reveal their possibilities in a multivariate context, we limit our univariate examination to some selected displays. Table 6.2 lists many of the graphical functions available in S-PLUS.

Table 6.2. Univariate graphical data exploration functions

S-PLUS Function	Description
boxplot	Boxplot display
density	Returns a density estimate for the distribution of the data. Use plot(density(x)) to see it graphically
dotchart	Dot chart
hist	Histogram display
identify	Identifies the data point next to the point clicked on in the graphics window
pie	Pie chart
plot	Standard plot, depending on the data
qqnorm	Quantile-quantile plot for the Normal distribution
qqplot	Quantile-quantile plot to check how well two data sets overlay
scatter.smooth	Plot of the data plus a smoothed regression line

Note: Most of the graphics functions have a long list of typically optional parameters. It is of course possible to use the functions only in their elementary form (like pie(x)), but if you want to add colors, modify the labels, explode one of the pies in the pie chart, or have different breakpoints for the histogram, you need to know what you can do. The boxplot function, for example, currently has 27 optional parameters. Have a look at the documentation to learn more about all the possibilities offered.

Figure 6.1 shows a simple plot of the four variables in the Barley data set. We split the screen into a 2x2 layout and plot each variable.

```
> par (mfrow=c(2, 2))              # 2x2 layout
> plot (yield)
> plot (variety)
> plot (year)
> plot (site)¹
```

---

[1] The variable labels are actually too long to fit. They were modified by the following commands (we will discuss the use of control characters in Section 8.4):

```
> levels (barley[[4]])[1] <- "Grand\nRapids"
> levels (barley[[4]])[3] <- "Univ. Farm"
```

Note that the `plot` function behaves differently for different types of data. The yield data are simply plotted against an index vector from 1 to 120 (the number of observations), and the factor variables `variety`, `year`, and `site` get a bar for each of their possible values. The length of the bar is determined by the number of values falling into this category. In this case, all values in the three variables have the same number of observations, showing again that we are dealing with a designed experiment - each combination of variables occurs exactly once.

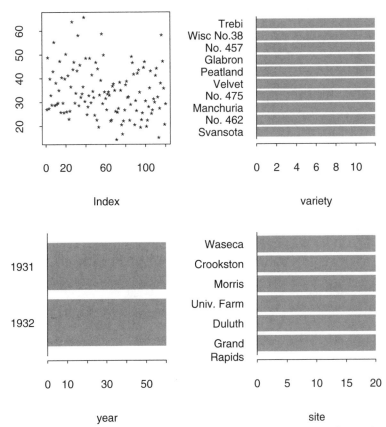

Figure 6.1. A display of the four variables `yield`, `variety`, `year`, and `site` in the `barley` data set.

Investigating the yields for the different sites by year a little further, we create histograms for each combination of `site` and `year`. As there are six sites in the data sets, each having two yields for 1931 and 1932, we obtain 6x2=12 histograms. They show ten values for the ten different varieties planted.

Figure 6.2 shows the histogram display by year and site.

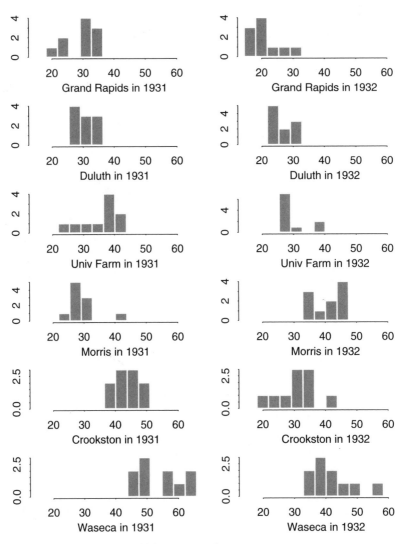

Figure 6.2. The barley yield by site and year.

The figure was obtained by entering

```
> par (mfcol=c(6, 2))                 # multiple figure layout
> yrange <- range (yield)
> limits <- seq (min (yield), max (yield), length=15)
> hist (yield[site=="Grand Rapids" & year==1931],
+ xlim=yrange, breaks=limits, xlab="Grand Rapids in 1931")
```

```
> hist (yield[site=="Duluth" & year==1931], xlim=yrange,
+ breaks=limits, xlab="Duluth in 1931")
> hist (yield[site=="Univ Farm" & year==1931],
+ xlim=yrange, breaks=limits, xlab="Univ Farm in 1931")
> ...
# etc. for the other subsets ²
```

In this way, we are able to review our data set in a concise form. The histograms can be compared column-wise, observing the yields for the different sites in 1931 and 1932, respectively. If they are viewed row-wise, we can see the different yields for the two years for a given site. More on this follows in Section 6.2.2 about multivariate graphical data analysis.

6.2 Multivariate Data Exploration

In the following section we will analyze the relationships between variables by examining them together or by conditioning one variable on the value of another. We saw how the barley yield differs when we condition on the year 1931 only, or on 1932. In this section, we will explore the relationship in more detail.

Note that we focus on data exploration by using exploratory techniques. Deterministic techniques like multivariate scaling and principal components analysis are left aside for the moment. If you know how these methods work, it will not be difficult to find the appropriate S-PLUS functions in the manual and apply them to a specific data set.

So let's continue examining the Barley data.

6.2.1 Descriptive Statistics

As before in the case of a univariate setting, let us put together some descriptive measures. Table 6.3 shows functions useful for multivariate data. The functions in Table 6.3 are used in the following analyses.

To see how a variable is distributed in two dimensions, S-PLUS offers the `table` and the `hist2d` function. We can determine how often different combinations of `year` and `site` occur by entering

---

² To be honest, we did not type in these long commands. If you know how to use `for` loops and `paste`, then you might understand how this works. If not, wait until Section 8.1.

```
> xrange <- range (yield)
> limits <- seq (min (yield), max (yield), length=15)
> for (ye in rev (levels (year)))
+   for (si in levels (site))
+     hist (yield[site==si & year==ye], xlim=xrange,
+       breaks=limits, xlab="", xlab=paste(si, "in", ye))
```

Table 6.3. Multivariate descriptive statistics functions

S-Plus Function	Description
hist2d	Tabulation of two-dimensional continuous data by counting the number of occurrences in intervals
table	Tabulation of discrete data of any dimension. Counts the number of data equal to the cell values
crosstabs	Creation of a contingency table from factor data
by	Applying a function to a data set by conditioning on the values of indices
split	Split up a data set and get a list with one component per group value
var	Calculate the variance or the covariance, if a matrix with more than one column is supplied

```
> table (year, site)
        Grand              Univ
        Rapids  Duluth  Farm  Morris  Crookston  Waseca
  1932      10      10    10      10         10      10
  1931      10      10    10      10         10      10
```

We see that every site had ten plantings in each year. If we look at the three-dimensional table of year, site, and variety, we can see that every combination of year, site, and variety occurs exactly once, as we have ten different varieties. We shorten the output because of its excessive length.

```
> table (year, site, variety)
  ,,Svansota
        Grand              Univ
        Rapids  Duluth  Farm  Morris  Crookston  Waseca
  1932       1       1     1       1          1       1
  1931       1       1     1       1          1       1

  ,,No. 462
        Grand              Univ
        Rapids  Duluth  Farm  Morris  Crookston  Waseca
  1932       1       1     1       1          1       1
  1931       1       1     1       1          1       1
  . . .
```

S-Plus shows the third dimension by "walking through the dimensions." It keeps the first value of the third dimension's variable fixed (in this case, the Svansota variety), and shows a two-dimensional table for all data having Svansota as its variety. The next table is shown for the second value for variety, here No. 462, and so on, until all third dimension values are taken.

To tabulate a variable like `yield`, which is a metric variable, we would not want to use `table`, because every observation would get a category of its own, as all the values are different. We could round the values and use `table`, or we can categorize them with the `hist2d` function. The `hist2d` function takes the data and categorizes them into intervals, just as the `hist` function does, but without graphical display. Because of the lack of two metric variables in the `barley` data, we tabulate `yield` against itself. The output consists of five elements; the midpoints of the x- and y-categories, the table containing the counts of the data, and two vectors giving the interval limits.

```
> hist2d (yield, yield)

$x:
[1] 15 25 35 45 55 65

$y:
[1] 15 25 35 45 55 65

$z:
          10 to 20  20 to 30  30 to 40  40 to 50  50 to 60  60 to 70
10 to 20       6         0         0         0         0         0
20 to 30       0        42         0         0         0         0
30 to 40       0         0        39         0         0         0
40 to 50       0         0         0        26         0         0
50 to 60       0         0         0         0         5         0
60 to 70       0         0         0         0         0         2

$xbreaks:
[1] 10 20 30 40 50 60 70

$ybreaks:
[1] 10 20 30 40 50 60 70
```

S-PLUS has informed us about the table we wanted to produce, and it stores the class breaks for the two variables in the elements `xbreaks` and `ybreaks`. Because of the leading $ sign, we can see that the returned data is stored in a list.

If we have a mixture of continuous and categorical variables and want to tabulate them, we can build classes on our own by rounding the metric variable. Let us round the yield to the nearest factor of ten. The `round` function has, besides the data, a second argument `digits`, telling us the number of digits to the right of the comma. If `digits` is negative, it refers to the number of digits to round off *to the left* of the comma. Rounding to the nearest factor of ten means that we need to set `digits` to -1.

```
> yield.round <- round (yield, -1)
```

We can check the result by tabulating our new variable, `yield.round`, which contains only the values 10, 20, 30, 40, 50, 60, and 70.

```
> table (yield.round)
   10   20   30   40   50   60   70
    1   18   51   31   13    5    1
```

Now we can tabulate variety against yield.round. You might want to see what happens if you tabulate variety against the original variable yield - without rounding the values - using the table function.

```
> table (variety, yield.round)
                  10   20   30   40   50   60   70
     Svansota      0    3    4    4    1    0    0
     No. 462       0    3    4    2    2    0    1
   Manchuria       0    2    8    1    1    0    0
     No. 475       0    4    4    3    1    0    0
      Velvet       0    2    5    4    1    0    0
    Peatland       0    0    8    3    1    0    0
     Glabron       1    0    5    5    0    1    0
     No. 457       0    2    5    3    1    1    0
  Wisc No. 38      0    1    4    3    2    2    0
       Trebi       0    1    4    3    3    1    0
```

We can see that most of the yields are around 30 bushels per acre. There is only a single harvest in the category 70, stemming from the species No. 462, but the species Wisc. No. 38 and Trebi have mostly high and only a few low yields.

Using the by function, we can investigate this a little further. Let us calculate a summary for each variety by using the function summary.

```
> by (yield, variety, summary)

INDICES:Svansota
         x
Min.    :16.63
1st Qu.:24.83
Median :28.55
Mean    :30.38
3rd Qu.:35.97
Max.    :47.33
------------------------------
INDICES:No. 462
         x
Min.    :19.90
1st Qu.:25.41
Median :30.45
Mean    :35.38
3rd Qu.:45.28
Max.    :65.77
```

```
------------------------------

...
------------------------------
INDICES:Wisconsin No. 38
         x
 Min.   :20.67
 1st Qu.:31.07
 Median :36.95
 Mean   :39.39
 3rd Qu.:47.84
 Max.   :58.80
------------------------------
INDICES:Trebi
         x
 Min.   :20.63
 1st Qu.:30.39
 Median :39.20
 Mean   :39.40
 3rd Qu.:46.71
 Max.   :63.83
```

It seems that Trebi and Wisconsin are the more fruitful varieties, as all figures are high in comparison to the other varieties.

If we wanted to split our data set into parts, for example to analyze each site on its own, we could select only a single variety by entering

```
> barley.Svansota <- barley [variety=="Svansota",]
```

to select only the Svansota data, or we could split up the data by using the split function.

```
> barley.split.by.variety <- split (barley, variety)
```

The first component of the list `barley.split.by.variety` looks like this now:

```
> barley.split.by.variety$Svansota
```

	yield	variety	year	site
13	35.13333	Svansota	1931	Univ Farm
14	47.33333	Svansota	1931	Waseca
15	25.76667	Svansota	1931	Morris
16	40.46667	Svansota	1931	Crookston
17	29.66667	Svansota	1931	Grand Rapids
18	25.70000	Svansota	1931	Duluth
73	27.43334	Svansota	1932	Univ Farm
74	38.50000	Svansota	1932	Waseca
75	35.03333	Svansota	1932	Morris
76	20.63333	Svansota	1932	Crookston
77	16.63333	Svansota	1932	Grand Rapids
78	22.23333	Svansota	1932	Duluth

We have all the Barley data of the Svansota variety in a separate structure. The variable `barley.split.by.variety` contains a list where the element names are the different varieties. If you enter

```
> names (barley.split.by.variety)
```

you will see the names of the list components. We see that for Svansota, for example, 1931 was a much better year than 1932, and that Waseca, Crookston, and University Farm had their biggest harvest in 1931, whereas the biggest harvests in 1932 were in Waseca, Morris, and University Farm.

6.2.2 Graphical Exploration

This section covers different methods of displaying the data. We present a variety of graphs to analyze the `barley` data set as well as the `geyser` data set. Exploring the data means displaying them in many different ways. The goal is to detect interesting aspects. New aspects can be examined further by means of other methods or by getting back to the sampling process and the sources of the data.

We start with an overview of what is available in S-PLUS. Table 6.4 presents graphics routines that can handle more than a single variable.

Table 6.4. Multivariate graphical data exploration functions

S-PLUS Function	Description
barplot	Barplot (bars with sub-bars stacked)
biplot	Plotting principal components and factor analysis results, show data and variables in a two-dimensional coordinate system
boxplot	Boxplot(s) of one or more variables in a single graph
contour	Contour lines of a two-dimensional distribution to access the common distribution function
coplot	Plot matrix of two variables conditioned on the values of a third variable
dotchart	Dotchart (values on parallel lines stacked) for a vector with optional variable grouping
faces	Chernoff faces illustrating by means of face elements a high-dimensional data set (up to fifteen variables)
hexbin	Hexagonal binning, a display technique for spatial data using hexagonally shaped bins
image	Image plots, colour scales for (geographical) heights or density values on a two-dimensional grid
matplot	Plotting columns of matrices against other matrix columns by using different symbols for each combination
pairs	Two-dimensional x-y scatterplots in a matrix of scatterplots for each combination of variables
persp	Perspective three-dimensional surface plots on a grid for density estimates
stars	Star plots of multivariate data in which each observation is displayed as a star. Each point is further away from the middle the larger the value is
trellis	Trellis displays for multivariate data, displaying any type of plot by conditioning on the remaining variables

Let us now graphically compare the sites' yields for the two harvests in 1931 and 1932. We choose the `boxplot` display to show all sites for a specific year in the same plot.

By reading the online documentation or the manuals, we find that if we want to display more than one data set in the same picture, the `boxplot` function expects a list of variables to plot. What we need to do is create a list containing one vector of yields for each of the sites. We have learned that the S-PLUS function `split` splits up a vector according to another variable, which we are going to use now. We split the `yield` by `site`, as we have done before, and supply the outcome to the `boxplot` function. Since we want to produce two plots, one for 1931 and one for 1932, we extract the data for the year first, then split up the extracted data.

```
> is.1931 <- year==1931              # True or False for 1931/1932
> boxplot (split (yield[ is.1931], site[ is.1931]),
+ main="Year 1931", ylim=range (yield))
> boxplot (split (yield[!is.1931], site[!is.1931]),
+ main="Year 1932", ylim=range (yield))
```

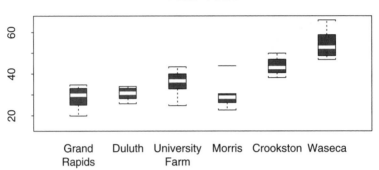

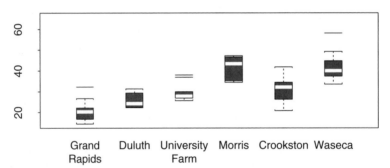

Figure 6.3. A boxplot display of the barley yield for all sites.

The result obtained is shown in Figure 6.3. Note the extraction of data by using the variable is.1931. The result is true for all data for which year is equal to 1931, and false otherwise (for all 1932 observations). In the first boxplot call, we select the data for which is.1931 is true, in the second boxplot we select the data for which it is not true.

If you are not familiar with boxplot displays, a box consists of a few basic elements: the lower quartile, the upper quartile, and the median (the dash in the middle). The box position tells us a lot about the data spread within a site and gives a comparison between the sites. The whiskers show the interval of values outside the box, and values far outside are represented by horizontal dashes. To learn more about this standard display tool, you might want to study Tukey (1977) or other books covering descriptive statistics or exploratory data analysis.

Many of the details in the preceding displays can be found again in this display. We see immediately that the Waseca site has very high yields, but the yields cover a much broader range than at Duluth, for example. Grand Rapids is always pretty low, and Morris is somewhat strange, as its yield in 1931 is very low, and in 1932, it is very high, in contrast to the other sites. For the other sites, the ordering is very stable for both years if ordered by yield size.

If you examine the display, you might discover other interesting details. It is also a good idea to have both plots on the same range for the y-axis, and right next to each other instead of on top of each other. Can you rearrange the graph in this way?

$\boxed{\text{Note}}$ If you try out the short example below, you will realize that there is a slight disturbance with the boxplot labels. The site names are sometimes too long, like "Grand Rapids" or "University Farm," and they overlap. What we can do in this case is to split the label into two lines by taking the result of split, a list, storing it into a variable, and renaming some labels. For example, we can make "Grand Rapids" a two liner by inserting the newline control character \n in the middle. The command would look like this.

```
> yield.split.1931 <- split (yield[is.1931], site[is.1931])
> names (yield.split.1931)          # to find "Grand Rapids"
> names (yield.split.1931)[[1]] <- "Grand\nRapids"
> boxplot (yield.split.1931)
```

We will learn more about control characters later on. They are summarized in Table 8.2 on page 179.   ◁

Displaying the boxplots in a different arrangement gives us some more information. We now use the same technique of splitting the data to display the yields for different sites. Figure 6.4 clearly reveals that the Morris site is the only one that had a lower yield in 1931 than in 1932. For all other sites, the box on the righthand side is clearly higher than the one on the lefthand side, indicating a higher yield.

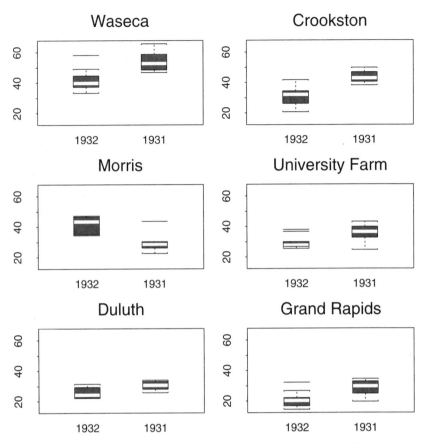

Figure 6.4. The Barley data in boxplot displays, categorized by site.

We continue examining the data by using Trellis displays, introduced by Cleveland (1993).

A Trellis display shows a matrix of plots. Each plot shows a graph of one variable against another, conditioned on the values of further variables. The plot in the position (i,j) of the matrix of plots contains only the values where the "row" variable takes its i-th value, and the "column" variable takes its j-th value.

Figure 6.5 shows such a Trellis display, and we see that the rows are determined by the year (the first row of the matrix shows only the 1931 data), and the columns are determined by the site (every column contains only values for the same site). Having split up the data like this, we know that the second plot in the first row shows the yields for all varieties for the year 1931 and site Duluth. How does this help us? We can easily compare a specific sites' yield for the different years by looking at the column of interest.

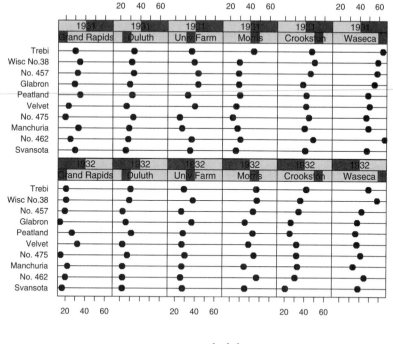

yield

Figure 6.5. A Trellis dotplot display of the Barley data.

On the other hand, we can compare the yields for both years by comparing the two rows. In other words, this graph shows us four variables at a time.

Producing a Trellis display is, in its basic form, very simple. Look at the display obtained in Figure 6.5 and compare it to the statement we used to produce it:

```
> dotplot (variety ~ yield | site + year, data=barley)
```

You see which variable goes where just by looking at the plot. To be more formal, we could add that the lefthand side of the formula (`variety ~ yield`) determines the inner plots, or the relationship we want to show. The vertical dash (`|`) separates it from the conditioning variables for the rows and columns of the matrix. In our case, the effect of `site+year` is the conditioning expression.

Note that this display is only one of many possible displays. The graphs in the matrix can take any form. If you wanted it to be a boxplot or a histogram of the current combination of conditioning variables, no problem. If, in addition, you were interested in conditioning on year and variety and looking at the yield at different sites in a graph, just rearrange the S-PLUS expression above. The S-PLUS documentation, especially the manual, provides lots of detailed examples. You might already have an idea of what could be done with other data sets you came across.

Looking once more at the call to the Trellis routine shows that we did not call a Trellis routine, but a `dotplot`. A dotplot is what we see in a single picture. Imagine that we wanted to modify the display of the dotplots. For this it is helpful to know that a graph within a Trellis graph matrix is produced by a *panel* function. These panel functions are typically named according to what they do. Our dotplot panel is stored under `panel.dotplot`. If you determine where the Trellis library is in your search path, you could determine the panel functions available (or consult the documentation). If the Trellis library were to be found in position 8 of our search path, we could determine what is available by listing all panel functions.

```
> objects (pattern="panel*", where=8)
```

Instead of writing our own panel function, we can simply copy one of the existing ones and modify it as needed. We take the function `panel.dotplot` and add a vertical line to the graphs where the mean of the current data is. Note that the panel function works on data that has already been subset to what the current picture displays. We just added the last line to the function and stored it under a new name, `new.panel`.

```
new.panel <- function (x, y, pch=dot.symbol$pch,
      col=dot.symbol$col, cex=dot.symbol$cex,
      font=dot.symbol$font, ...)
{  ok <- !is.na(x) & !is.na(y)
   dot.symbol <- trellis.par.get ("dot.symbol")
   dot.line <- trellis.par.get ("dot.line")
   abline (h=y[ok], lwd=dot.line$lwd,
     lty=dot.line$lty, col=dot.line$col)
   points (x, y, pch=pch, col=col, cex=cex, font=font, ...)
   abline (v=mean(x), lwd=3, col=3)
}
```

After having modified the panel function, we call `dotplot` again to produce a Trellis graph using our new panel function.

```
> dotplot (variety~yield | site+year, data=barley,
+ panel=new.panel)
```

The result appears in Figure 6.6.

As we suspected before, the mean value for the yield is again higher in 1931 for all sites except Morris.

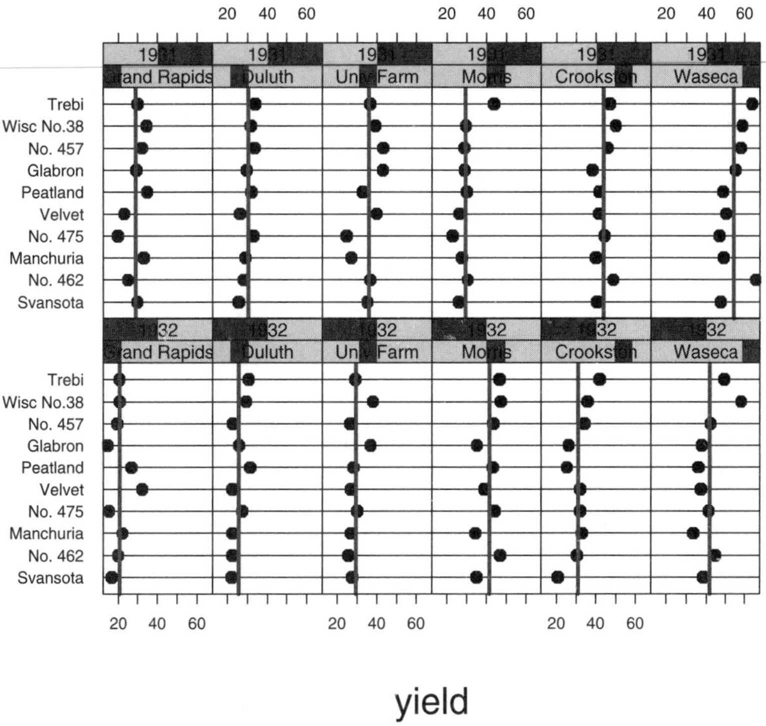

Figure 6.6. A customized Trellis display of the Barley data.

Using these Trellis displays, Cleveland (1993) and Becker, Cleveland, and Shyu (1996) conclude that the years 1931 and 1932 must have been swapped for Morris. Of course this can no longer be proven, but there is some clear evidence for it. Although this data set has undergone many analyses in the statistical literature, this obscureness had not been detected, which shows how graphical methods can help provide more insight into a data set.

To illustrate more graphical techniques, we examine the Geyser data set in detail. This data set was previously introduced in Section 3.6.

The Geyser Data Set. The Geyser data consist of continuous measurements of the eruption length and the waiting time between two eruptions of the Old Faithful Geyser in Yellowstone National Park in 1985 in minutes. Some duration measurements, taken at night, were originally recorded as S (short), M (medium), and L (long). These values have been coded as 2, 3, and 4 minutes, respectively. The original publication by Azzalini and Bowman (1990) provides more details.

Having a look at the `geyser` data (have a look!) shows that they are stored in a list with two components, `waiting` and `duration`. Lists can be attached, such that we can use `waiting` instead of typing `geyser$waiting` all the time.

```
> attach (geyser)
```

We might want to begin by examining the basic data characteristics.

```
> summary (waiting)
      Min.  1st Qu.  Median   Mean  3rd Qu.   Max.
       43       59      76  72.31      83    108
> summary (duration)
      Min.  1st Qu.  Median   Mean  3rd Qu.   Max.
    0.8333       2       4  3.461    4.383   5.45
```

This gives us a basic impression. The duration of an eruption lies between less than a minute and almost 6 minutes, and the waiting time for the next eruption lies between a little more than 40 minutes and more than 100 minutes. On average, if you just missed the last eruption, then you would have to wait for more than an hour to see an eruption of approximately 4 minutes.

Next we are interested in a graphical data display. Figure 6.7 shows what the data look like.

From this graph, we can already see some surprising details. We already know that parts of the data were set to 2, 3, and 4 minutes, such that quite a few points lie on these lines parallel to the x-axis. Secondly, it seems that the data also lie on lines parallel to the y-axis. Examining the data confirms that `waiting` time was measured in integer minutes.

Next, we see three clusters of point clouds. Interestingly enough, they are more or less clearly separable.

We can use an interactive facility of S-PLUS to determine the cluster's boundaries. Plot the data and use the mouse to click on some points in the graph. Calling the function `locator`, S-PLUS will return the coordinates of the points you click on with the left mouse button. Tell S-PLUS to stop recording the clicks by clicking on the right mouse button.

```
> plot (waiting, duration)
> locator()
```

If we cut the x-axis at x=67 and the y-axis at y=3.1, the plot is divided into four areas, three of which contain a cluster of data points, and the fourth,

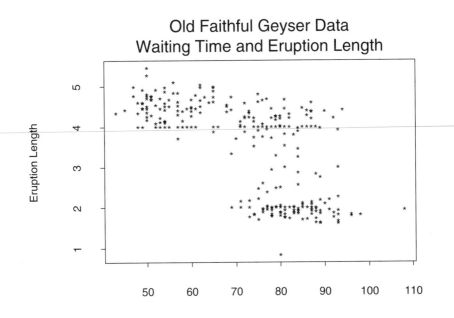

Figure 6.7. The Old Faithful Geyser data.

the one in the lower left position, contains no data at all. This might require questioning the Park Ranger who collected the data.

To show it to the ranger, we plot the graph again and add the dividing lines to it. Furthermore, we choose a different plotting character for each of the four subsets, and label the extreme value in the lower right corner (the only one with a waiting time of more than 100 minutes).

The following commands produced Figure 6.8. We first plot a graph *with no points*, which contains everything except for the points (the axes, labels, and other layout parameters). We then add the horizontal and vertical line and finally, the points.

```
> plot (x, y, type="n", ylab="Eruption Length",
+ xlab="Waiting Time for the Eruption")
> title ("Old Faithful Geyser Data\nWaiting
+ Time and Eruption Length")
> abline (h=3.1)
> abline (v=67)
```

The data are divided into four subcategories, which we determine by the logical expressions subset1, ..., subset4. These vectors of logical TRUE/FALSE elements determine whether or not a point belongs to the subset.

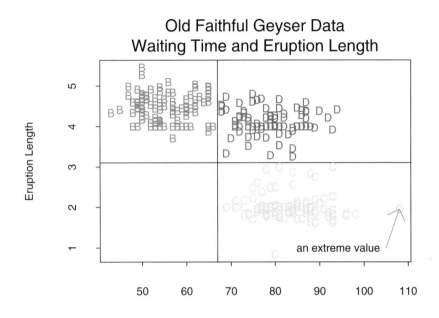

Figure 6.8. The Old Faithful Geyser data in subsets.

```
> subset1 <- (waiting < 67) & (duration < 3.1)
> subset2 <- (waiting < 67) & (duration > 3.1)
> subset3 <- (waiting > 67) & (duration < 3.1)
> subset4 <- (waiting > 67) & (duration > 3.1)
```

The next step is to plot the four groups separately by using the **points** function, but displaying different characters (instead of points) and using different colors to demarcate the groups.

```
> points(waiting[subset1],duration[subset1],pch="A",col=1)
> points(waiting[subset2],duration[subset2],pch="B",col=2)
> points(waiting[subset3],duration[subset3],pch="C",col=4)
> points(waiting[subset4],duration[subset4],pch="D",col=8)
```

Finally, we add an arrow pointing to the extreme point with a waiting time of about 110 minutes. Note that we do not determine the end point of the arrow, but go to its waiting time (which we know is the maximum value) and the corresponding duration (by selecting the index of the maximum waiting time). We want a v-shaped arrow; therefore, **open=T** is set. Lastly, we place the text "an extreme value" on the bottom line of the plot, ending where the arrow starts (**adj=1** means the text is right-justified). The text we want to add has two extra spaces at the end, such that there is a little gap between the text and the arrow's end point.

```
> arrows (105, 1, max (waiting),
+ duration[waiting==max (waiting)], open=T)
> text (105, 1, "an extreme value ", adj=1)
```

We have already seen that the function hist2d is useful for putting continuous data like the Geyser data into a table frame. If we do it with our Geyser data set, we get back a list with five components, telling us the mid-values of the chosen intervals and the end points, together with the tabulated data.

```
> hist2d (waiting, duration)
```

```
$x:
[1]   45   55   65   75   85   95 105
```

```
$y:
[1] 0.5 1.5 2.5 3.5 4.5 5.5
```

```
$z:
```

	0 to 1	1 to 2	2 to 3	3 to 4	4 to 5	5 to 6
40 to 50	0	0	0	0	16	0
50 to 60	0	0	0	1	56	3
60 to 70	0	0	1	2	28	1
70 to 80	0	13	19	9	40	0
80 to 90	1	31	25	9	24	0
90 to 100	0	11	3	2	3	0
100 to 110	0	1	0	0	0	0

```
$xbreaks:
[1]   40   50   60   70   80   90 100 110
```

```
$ybreaks:
[1] 0 1 2 3 4 5 6
```

The table is put into the component z, and we see that the visibility of effects depends heavily on the intervals chosen. The effect of the three point clouds is not that clearly visible, although you might guess at them by studying the values. Experimenting with different choices of interval bounds might give more insights.

You can also use the hist2d function that was introduced on page 93 to set up data to be used as input to other functions. The persp function produces a 3D type plot that accepts input from hist2d.

```
> persp (hist2d (waiting, duration))
```

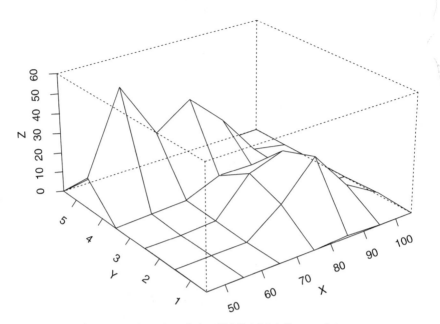

Figure 6.9. A perspective plot of the Old Faithful Geyser data.

We create Figure 6.9 showing a perspective plot of the surface of the empirical distribution. The x- and y-axis are set up by taking the variables xbreaks and ybreaks of the hist2d output, and the matrix of values in the z component determines surface height. Try to reveal the three peaks more clearly by choosing another set of intervals for hist2d to display the perspective surface plot.

You can use the same data from hist2d to get an image display of the surface. The image display and the corresponding S-Plus function image are often used to display geographic data, using latitude and longitude as axes, and the geographical height of the area is color-coded. This is the principle of every contour map. We use the color display to look at the tabulated Geyser data "from the top," getting the heights in different colors.

```
> image (hist2d (waiting, duration,
+ xbreaks=seq (40, 110, by=5), ybreaks=seq (0, 6, by=0.5)))
> title ("An image plot of the geyser data")
```

For a nicer display, we chose different classes than the default settings used above. The result is shown in Figure 6.10, and although the loss of the coloring in this picture is a disadvantage, we can still see the three "mountains" as well as the "desert area" in the lower left corner. Other options available are different color schemes, like a "heat color scheme" ranging from cold (blue) to hot (red).

An image plot of the geyser data

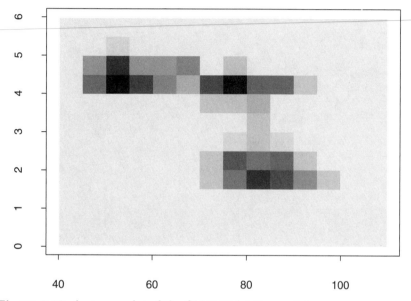

Figure 6.10. An image plot of the Old Faithful Geyser data.

It becomes obvious that we need to experiment with parameters if we examine Figure 6.11. We show two pictures of the same data, but the underlying categorizations into tables are different. From the picture on the lefthand side we see the three groups in the data, whereas the picture on the righthand side leads us to conclude that we have only two mountain tops.

This figure shows contour lines, or in other words, height lines of a data set. The `contour` function is used to create Figure 6.11, which also accepts input from `hist2d`.

```
> contour (hist2d (waiting, duration))
> contour (hist2d (waiting, duration,
+ xbreaks=seq (40, 110, by=5), ybreaks=seq (0, 6, by=1.5)))
> title ("A contour plot of the geyser data")
```

This excursion into data analysis using graphics for exploring structure reveals that the term "exploratory data analysis" describes a search for effects discovered by using many different displays and summaries of the same data set. *A single number or display can never describe the whole complexity of a data set.*

A contour plot of the geyser data

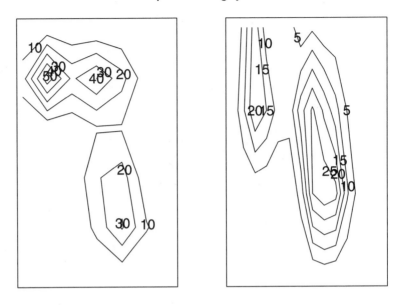

Figure 6.11. A contour plot of the Old Faithful Geyser data.

If you would like to explore the Barley or the Geyser data set with inter-active techniques and rotating point clouds, try out the functions referenced in Table 6.5. Use the mouse to define a brush and move it over point clouds to mark ("highlight") them. If you get stuck, consult the manuals. Just do it.

Table 6.5. Multivariate dynamic graphics functions

S-PLUS Function	Description
brush	Interactive marking of subsets on 2D scatterplots
spin	Rotating data, marking and highlighting of subsets

6.3 Distributions and Related Functions

This section examines a variety of distributions covered by built-in S-PLUS functions. As we will see later on, adding further distributions to the existing set is also very straightforward. We divide the distributions themselves into continuous and discrete distributions. Presently, all these distributions are univariate, meaning that the spaces of the densities are one-dimensional.

6.3.1 Univariate Distributions

Statistical distributions occur in many practical data analysis issues. Most models to describe data behavior are based on distributional assumptions, and derive the estimates for unknown model parameters from the underlying distributions.

Very often, simulations are conducted to check accuracy, robustness, and sensitivity. For these purposes, random number generators are used.

The most important functions related to a distribution are

- the density function, which specifies a random variate distribution (with fixed parameters).
- the probability function, which is the integral over the distribution function, or, in the discrete case, the sum up to the specified point. In other words, the probability function at x gives the probability of the random variable X being less than or equal to x.
- the quantile function, which is the inverse of the probability function. If p is the probability function and q the quantile function of a continuous distribution, then p(q(x))=q(p(x))=x for any x.
- the random number generating function, which generates one (or more) numbers distributed according to the specified distribution function. For a large set of random numbers, the distribution of the random numbers (visualized, for example, by a histogram), looks approximately like the density function.

S-PLUS has a systematic naming of functions related to distributions. The functions related to the same distribution have the same name except for the first letter. The first letter indicates what type of function it is. Table 6.6 explains the system.

Table 6.6. Categorization of distribution-related functions in S-PLUS

Type[1] Character	Function Type
d	Distribution function
p	Probability function or cumulated density function
q	Quantile function, the inverse of the probability function
r	Random number generation

[1] The significant letter for identifying the functionality of the distribution function is either d, p, q, or r. See also Table 6.7.

All these functions take numbers and vectors as input arguments. For a vector, the corresponding value for each element is computed. Then, to identify a specific function, we need to know the abbreviations for the distributions available to us, which are listed in Table 6.7.

Table 6.7. List of distribution-related functions in S-PLUS

Distribution	S-PLUS Abbreviation[1]	Parameters[2]	
	Continuous Distributions		
Beta	beta	shape1	shape2
Cauchy	cauchy	location=0	shape=1
Chi-square	chisq	df	
Exponential	exp	rate=1	
F	f	df1	df2
Gamma	gamma	shape	
Logistic	logis	location=0	scale=1
Lognormal	lnorm	meanlog=0	sdlog=1
Normal	norm	mean=0	sd=1
Stable	stab	index	skewness=0
Student's t	t	df	
Uniform	unif	min=0	max=1
Weibull	weibull	shape	
	Discrete Distributions		
Binomial	binom	size	prob
Geometric	geom	prob	
Hypergeometric	hyper	m, n, k	
Negative Binomial	nbinom	size	prob
Poisson	pois	lambda	
Discrete Uniform	sample	x, size=n, replace=T	
Wilcoxon Rank Sum	wilcox	m	n

[1] The characteristic letter, one of d, p, q, r, plus the listed abbreviation, gives the name of the S-PLUS function.
[2] If parameters are preset with a default value, like in mean=0, they do not need to be specified unless another parameter value than the default is desired.

Tables 6.6 and 6.7 list the function types available. To determine the function's name, combine the characteristic letter from Table 6.6 with the abbreviation in Table 6.7. For example, to generate random numbers from the Normal distribution, the function to use is rnorm ("r" + "norm").

There are four functions related to the Normal distribution,

pdf — dnorm for calculating the value of the distribution function,
cdf — pnorm for calculating the value of the probability function,
qnorm for calculating the inverse probability function, and
rnorm for generating random numbers from the Normal distribution.

These four functions all have the same parameters, as they refer to the same distribution, namely **mean** and **sd**, the mean and the standard deviation. Both arguments are optional. They are preset to define the standard Normal distribution with mean 0 and standard deviation 1.

| **Note** | The standard deviation, not the variance is the argument used for dispersion by the functions for the Normal distribution. The standard deviation is simply the square root of the variance.   ◁

You can check out to see the numerical accuracy of the distributional functions. Calculate the difference between some values x and p(q(x)) or q(p(x)), which should be 0 in theory (p stands for the probability and q for the quantile function of the distribution).

| **Note** | As you might know, the random numbers generated by a computer are never really random. They consist of a sequence of numbers, where each number depends somehow on the previous number generated. S-PLUS uses the variable .Random.seed to store the state of the random number generator. You can use set.seed(n), where n is any number, to produce the same sequence of random numbers several times, maybe to test a self-written routine.   ◁

Graphing Distributions

You will often want to see what a distribution looks like, which usually means looking at the density function. Here we need to make a distinction between continuous and discrete distributions.

Recall that a continuous density has a continuous support; that is, all values of the density function in a given interval are greater than zero. This interval usually ranges from minus infinity to plus infinity. For this reason, we cannot graph the "whole" distribution from minus to plus infinity, but have to restrict ourselves to a part of it. A good idea is to plot, for example, 90% of the density support.

To calculate these 90% limits, we cut off 5% on each side and get the boundaries by calculating the 5% and 95% quantiles. As an example, we plot the standard Normal (0,1), t (5), and standard Cauchy distribution and calculate the boundaries as described.

```
> x <- c(0.05, 0.95)
> bounds.normal <- qnorm (x)
> bounds.t      <- qt (x, 5)
> bounds.cauchy <- qcauchy (x)
```

Next, we create a sequence of points at which we want to calculate the densities. Therefore, we calculate the common range of the boundaries of the three distributions.

```
> bounds <- range (bounds.normal, bounds.t, bounds.cauchy)
> points.x <- seq (bounds[1], bounds[2], length=1000)
```

Now the values of the density functions are easily obtained.

```
> points.normal <- dnorm (points.x)
> points.t       <- dt (points.x, 5)
> points.cauchy <- dcauchy (points.x)
```

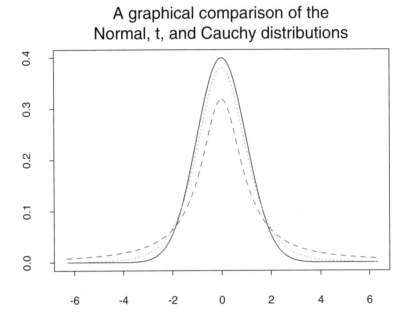

Figure 6.12. A graphical comparison of three distributions.

Plotting the functions is straightforward now. Be sure to calculate the maximum value of the y-axis data first, in order to have all values inside the plotting range. A trick is to plot the "most outward" curve first, the curve having the largest and smallest values. Then the other ones will fit into the given range. We calculate the boundaries first, plot an empty frame, and then add the three lines.

```
> plot (0, 0, type="n", xlim=bounds,
+ ylim=range (c(points.normal, points.t, points.cauchy)),
+ xlab="", ylab="Density Value")
> lines (points.x, points.normal, col=1, lty=1)
> lines (points.x, points.t,      col=2, lty=2)
> lines (points.x, points.cauchy, col=3, lty=3)
```

We add a title and obtain a result like in Figure 6.12.

```
> title ("A graphical comparison of the\n
+ Normal, t, and Cauchy distributions")
```

What we see is that the Normal distribution is more peaked at x=0 than the
other two distributions and has shorter tails. For a discrete distribution, the
density can only be calculated at discrete points. For example, consider the
distribution function of the Binomial distribution - the function describing
the probability of x heads out of n coin tosses, where the probability of a
head for a single toss is p. The distribution function is greater than zero only
at points 0, ..., n. To graph this adequately, we plot vertical height bars at
the discrete points. The result is shown in Figure 6.13.

Density of the Binomial (20,0.6) distribution

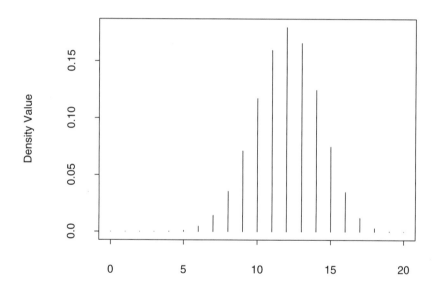

Figure 6.13. A graph of the Binomial distribution (n=20, p=0.6).

```
> n <- 20                              # number of coin tosses
> p <- 0.6                             # probability for heads
> x <- 0:n
> y <- dbinom (x, n, p)                # prob. for 0..n heads
> plot (x, y, type="h", xlab="", ylab="Density Value")
> title ("Density of the Binomial (20, 0.6) distribution")
```

6.3.2 Multivariate Distributions

Currently, S-PLUS does not cover multivariate (more than one-dimensional) distributions. Typically, most of the real world data sets are in fact multivariate, but there are not so many distributions used in standard data analysis. Mostly, the multivariate Normal (or Gaussian) distribution is used. Therefore, we refer briefly to the multivariate Normal. If you require other distributions not covered by the S-PLUS library, they are typically easy to implement if the corresponding function or the algorithm for generating random numbers are at hand. Textbooks like the four volumes of Johnson and Kotz (1969), Devroye (1986), or Ripley (1987) cover many of the distributions.

The Multivariate Normal Distribution

S-PLUS contains a function to generate a set of multivariate Normal random numbers. If you look at the help text on the keyword *normal,* you can simply copy it from there and paste it into your S-PLUS session.

Using this, we can directly write an S-PLUS function to generate multivariate Normal random numbers, which we call `rmultnorm`. A matrix stores the output, where each row represents one multivariate random number (vector). For example, calling

```
> mu <- c(1, 2, 5)
> Sigma <- matrix (c(1,2,3,2,3,4,3,4,5), 3, 3)
> x <- rmultnorm (100, mu, Sigma)
```

generates 100 random numbers, and the matrix elements `x[1,]`, `x[2,]`, etc. contain one realization each. We can check for the mean and the variance by calculating these from the sample:

```
> apply (x, 2, mean)          # the column-wise mean
> var (x)                      # the covariance matrix
```

6.4 Confirmatory Statistics and Hypothesis Testing

After learning to analyze data through exploratory methodology, and how distributions and their related functions work in S-PLUS, we can proceed toward a possible next step, confirmatory statistics. Confirmatory or classical statistics typically is the problem of having a hypothesis about a data set and an alternative. The problem consists of deciding in favor of the hypothesis or rejecting the hypothesis in favor of the alternative.

We will not discuss the details of testing statistical hypotheses here, as there is a wealth of textbooks available on this subject. In addition, the manuals provided with S-PLUS offer an excellent quick overview of the subject if you already have some basic knowledge. This section will simply show what could be done by presenting an example of using a Student t-test and having a look at the results.

Let us do a simple Student's t-test using the Barley data we examined before. The goal is to decide whether the average yield of one site is different from the average yield of another site. This gives us twenty observations for each site, and the sites to compare are Crookston and Morris.

As we assume a Normal distribution for the data, and we do not assume a known variance of the yield, we need to conduct a t-test. The significance level alpha (α) is set to 5% or 0.05. As we know, each observation is made in a specific year and for a specific variety at both of the sites, Crookston and Morris. Thus we can think of the data as correlated, or paired, since the same years and varieties were used. If this were not the case, we would have to use the more standard unpaired, or two-sample, Student's t-test. The Barley data are already sorted within site by year and variety, so we are ready to do our test. Note that the confidence level is 0.95, being equal to 1–0.05.

```
> t.test (yield[site=="Crookston"], yield[site=="Morris"],
+ paired=T, conf.level=0.95)

                    Paired t Test

    data:   yield[site == "Crookston"] and
            yield[site == "Morris"]
    t = 0.6595, df = 19, p-value = 0.5175
    alternative hypothesis:
        true difference in means is not equal to 0
    95 percent confidence interval:
    -4.390992  8.430988
    sample estimates:
    mean of x - y
        2.019998
```

What is the output? S-PLUS tells us that we calculated the paired t-test, and gives us the value of the t-distributed test statistic, 0.6595. Based on 19 degrees of freedom (20 sample pairs minus one degree), this results in a "p-value" of 0.5175. For a p-value smaller than alpha=0.05, we would have concluded that we should reject the hypothesis.

In our case however, we decide that we cannot reject the null hypothesis, as there is not enough evidence that the mean yields from Crookston and Morris differ from one another. This means that the estimated difference in mean yields of 2.02 is not statistically significant from zero. In fact, the 95% confidence interval of (-4.39, 8.43) indicates where we expect the true mean difference to lie. Note that the confidence interval contains the mean value of zero, i.e. no difference between the two sites.

Earlier on, we found some evidence that at the Morris site, the years 1931 and 1932 had been accidentally swapped. This makes wary of deciding to use a paired t-test. You can check on your own how big the difference is, if an *unpaired* t-test is carried out that does not assume related pairs of data. (A look at Figure 6.14 shows that the difference is not very large.)

Figure 6.14 shows the most important information in a graphical display. The shaded areas form the rejection area at the 5% level (two-sided), where the hypothesis of the mean being 0 is rejected in favor of the alternative. The middle range shows the acceptance region, where the hypothesis cannot be rejected, and the two values of the t-test for the paired and unpaired samples have been added. Note that we get different degrees of freedom for the two t-tests, 19 for the paired and 38 for the unpaired test, but the difference in the graphical display is almost invisible.

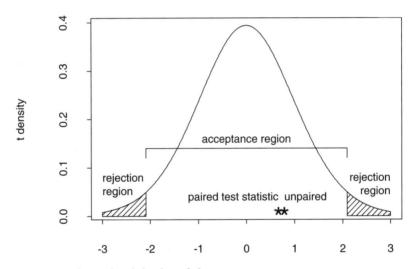

Figure 6.14. A graphical display of the t-test.

Refer to the manual for further details about testing, the background of testing, and the detailed usage of S-PLUS functions. If you are familiar with a test, studying the reference manual or the online documentation should provide the information you need to carry it out.

Finally, the following note helps to do the complete calculation of test statistics and critical values in S-PLUS. Looking up values in tables should be obsolete from now on.

| Note | If we calculate a test statistic and need to know the critical values of the test, we can calculate them directly with an S-PLUS statement, if the distribution is in the set of distributions provided with S-PLUS. If the test is

Table 6.8. Statistical tests in S-PLUS

S-PLUS Function	Description of the Test
binom.test	Exact binomial test
chisq.gof	Chi-square (χ^2) goodness of fit test
chisq.test	Chi-square (χ^2) test on a 2-dimensional contingency table
cor.test	Test for zero correlation of two samples
fisher.test	Fisher's exact test on a two-dimensional table
friedman.test	Friedman rank sum test
kruskal.test	Kruskal-Wallis rank sum test
ks.gof	Kolmogorov-Smirnov test for one or two samples
mantelhaen.test	Mantel-Haenszel test
mcnemar.test	McNemar test on a two-dimensional contingency table
prop.test	Test for proportions of success
t.test	Student's t-test for one and two samples
var.test	F test for comparison of variances of two samples
wilcox.test	Wilcoxon (Mann-Whitney) rank sum and signed rank test

one-sided, meaning that the alternative hypothesis is "greater than" or "less than," and the underlying distribution of the test statistic is, for example, a Normal distribution, the critical value is calculated as

```
> qnorm (1 - alpha, m, s)        # greater than
> qnorm (alpha, m, s)            # less than
```

where *alpha* is typically 0.05 or 0.01, and *m* and *s* are the mean and standard deviation of the hypothesis. If the test is two-sided, the critical values are the $\alpha/2$ and $1 - \alpha/2$ quantiles of the test statistic distribution. Taking the example of the Normal distribution again, we would calculate the following critical values.

```
> qnorm (c(alpha/2, 1-alpha/2), m, s)
```

◁

Creation of a Graph Illustrating a Statistical Test

The following describes how Figure 6.14 was created. It combines what we have learned about plotting distributions with graphical illustration of statistical testing. The combination of many functions shows how you can easily create even complex graphics like Figure 6.14. We store the two yield subsets in variables with shorter names, just for convenience. We calculate the degrees of freedom and set the significance level in a variable.

```
> x <- yield [site=="Crookston"]
> y <- yield [site=="Morris"]
> df <- length (x)-1           # paired df
> alpha <- 0.05
```

The next step sets up the graphical display. We set the graph's boundaries to -3 and 3 and store them in variables. We select 1000 equally spaced points to form the x-axis, and calculate the value of the t distribution for each of these points. These values are then plotted against each other.

```
> bound.left <- -3
> bound.right <- 3
> xaxis <- seq (bound.left, bound.right, length=1000)
> yaxis <- dt (xaxis, df)
> plot (xaxis, yaxis, type="l", xlab="", ylab="t density")
```

Now we plot the shaded areas of rejection. What we need is a polygon "walking" around the area we want to shade. We calculate the critical values and set up the x-axis from the left limit of the graph to the critical value, and the y-values are calculated again as the density values of the t distribution. We add two points, (critical.left, 0) and (bound.left, 0), to complete the "walk around the shaded area." Calling `polygon` with the optional parameter `density=25` adds a shaded area to the figure.

```
> critical.left <- qt (alpha/2, df)
> critical.right <- qt (1-alpha/2, df)
> xaxis <- seq (bound.left, critical.left, length=100)
> yaxis <- c(dt(xaxis, df), 0, 0)
> xaxis <- c(xaxis, critical.left, bound.left)
> polygon (xaxis, yaxis, density=25)
```

We use the same idea to draw the rejection region on the righthand side, just symmetric to the area on the left.

```
> xaxis <- seq (critical.right, bound.right, length=100)
> yaxis <- c(dt(xaxis, df), 0, 0)
> xaxis <- c(xaxis, bound.right, critical.right)
> polygon (xaxis, yaxis, density=25)
```

We calculate the test statistic for the paired and unpaired tests next, and access and store the t-statistics directly. See the manual pages to look at what is returned from `t.test`. Next, we plot the points of the test statistic and add text to it, once right-justified and once left-justified, such that we get a back-to-back effect.

```
> test.stat <- t.test (x, y, paired=T,
+   conf.level=1-alpha)$statistic   # paired test
> points (test.stat, 0.01, cex=1.5, pch="*")
> text (test.stat, 0.04, "paired test statistic", adj=1)

> test.stat <- t.test (x, y, paired=F,
+   conf.level=1-alpha)$statistic   # unpaired test
> points (test.stat, 0.01, cex=1.5, pch="*")
> text (test.stat, 0.04, "unpaired", adj=0)
```

Finally, we add the text for the rejection regions, adjusted to the borders, and draw the line showing the acceptance region's limits by creating a vector with four points, which covers the polygon line. On top of it, we add a string describing the interval line.

```
> text (bound.left , 0.08, "rejection\nregion", adj=0)
> text (bound.right, 0.08, "rejection\nregion", adj=1)
> text ((bound.left+bound.right)/2, 0.16,
+ "acceptance region")
> xaxis <- c(rep(critical.left, 2), rep(critical.right, 2))
> yaxis <- c(0.12, 0.14, 0.14, 0.12)
> lines (xaxis, yaxis)
```

The little program shows the basic elements of tests, critical values, and acceptance and rejection regions for those of you not so familiar with such ideas. As a bonus, you got some ideas of how flexible S-PLUS graphs are, and some tricks to use in your own graphs.

6.5 Missing Values

Missing values can originate from different sources. For example, in a survey of inhabitants of a city, the people asked simply did not answer some of the questions. The unanswered questions would be entered into the data set as "missing" values.

On the other hand, some undefined calculations can also produce a missing value. Try to calculate the logarithm of a negative number. S-PLUS will give a missing value as the result, together with a warning that the calculation generated some missing values. The warning is very helpful. If you are warned that missing values were generated after execution of a more complex function, it is a good indication that something probably went wrong.

The code for a missing value in S-PLUS is NA. The two letters NA form an abbreviation standing for *Not Available*. NA values have no associated type. They can occur in numeric data as well as in character data or any other structure. NAs can be entered from the command prompt, or they can occur in a data file. S-PLUS will recognize them as missing and treat them accordingly.

⸤Note⸣ **The general rule for any type of operation is: A missing value (NA) combined through an operation with any other data results in a missing value again.**   ◁

Following are some details about how to detect missing values and how to use them.

6.5.1 Testing for Missing Values

The function is.na checks if a value is an NA. It works elementwise, which
means that if a vector is supplied to is.na, a vector of TRUE/FALSE values
is generated with a TRUE occurring for each corresponding NA in the original
vector. The same holds for other structures like matrices or arrays.

```
> x <- log (-2:2)
> is.na (x)
        T T T F F
```

This function can then easily extract all non-NAs from a variable.

```
> x.no.na <- x [!is.na (x)]
```

If you no longer remember how the non-missing values are extracted, review
the introductory section on missing values (Section 4.2, page 49).

Supplying Data with Missing Values to Functions

Most S-PLUS functions do not perform their intended action by simply ig-
noring missing values, if there are any. However, this is frequently exactly
what you want to do.

That is, you often want to calculate the mean or variance over just the
non-missing data, but to assure that nothing unintended happens, you must
set a parameter to do it. By default, this option is turned off to avoid supplying missing values produced by another function (this is helpful in error-
tracing). The parameter is called na.rm.
Here is a short example:

```
> x <- c(1,2,NA,4)
> mean (x)
        NA
> mean (x, na.rm=T)
        2.333333
```

In the latter expression, the S-PLUS function removes all missing data and
performs the action on the remaining data. The result is that the mean
function divides the sum of data in the above example by three and not by
four, the total length of x.
The same effect is obtained by removing the missing data "by hand," like in

```
> mean (x[!is.na (x)])
```

6.5.2 Missing Values in Graphs

Missing values are ignored in graphical routines except for special cases like when you want to plot x, which consists only of missings (this would return an error message). So, for example, the `plot` and `lines` functions ignore all points where one of the coordinates is an `NA`.

| Note | Using NAs can be useful if you *want* to plot an interrupted line. Example:

```
> plot (1:10, c(2,3,2,4,NA,3,2,2,4,5), type="l")
```

creates a line, which is not connected between the points (4,4) and (6,3). ◁

Infinite Values

There is another category of special values, the infinite values. Typically, infinite values are not part of a data set, but are sometimes created from a calculation. For example, dividing a number by zero leads to an infinite value. But since the number you divide by zero can have a sign, there is also plus and minus infinity. Dividing a positive number by zero leads to the "value" plus infinity, and dividing a negative number by zero leads to minus infinity.

Infinity is displayed as `Inf`. Similar to missing values and the `NA` code, `Inf` can be entered from the keyboard or can be in a file. If S-PLUS comes across `Inf`, it interprets it as infinity. You can check for infinite values by using `is.inf`.

Sometimes it might be necessary to check for missing and infinite values separately, as an infinite value is not missing and vice versa. See the following example.

```
> x <- (-1:1)/0
> x
        -Inf NA Inf
> is.na (x)
      F T F
> is.inf (x)
      T F T
```

There are even some simple rules for infinite values. Try to add, subtract, or multiply infinite values with themselves and look at the results.

6.6 Exercises

Exercise 6.1

Generate 100 and 1000 random numbers, both samples from a Normal distribution with mean value 3 and variance 5. Draw histograms with bin width 0.5, 1, and 2 for each of the two samples.
Remember that all figures have exactly the same underlying distribution.
Plot them all in a single graphics window and label them accordingly. What is visible?

Exercise 6.2

Create a vector containing three real numbers, one missing value (NA), and one infinite value (Inf). Calculate the mean and variance of the non-missing data. Figure out how the missing and infinite data are treated in a simple plot and histogram. What happens if you divide x by itself? Can you explain this behavior?

Exercise 6.3

Do an analysis of the Car data set provided with S-PLUS. The data set, stored under car.all, contains a set of cars plus a variety of their attributes, like fuel consumption, price, and many more.
Before starting the analysis, determine which car models are contained in the data and what the variables given in the data set are. Extract the fuel consumption (miles per gallon) and determine for each type of car (small, medium, compact, large, van, sporty) the model with the lowest and the highest fuel consumption.
On the basis of average fuel consumption and tank size, calculate the maximum travel distance without a gas refill.
Take the variables maximum travel distance, tank size, mileage, and horsepower, and display them graphically. Try to determine, using only graphics, the two variables with the strongest dependency.
Check your guess by calculating the correlation matrix for these variables.

Exercise 6.4

Test the S-PLUS random number generator with the "dice test." Simulate a number of die rolls and check if they all occur with approximately the same frequency. In other words, we need to simulate a die roll, where the six possible outcomes (1, 2, 3, 4, 5, and 6) occur with the same probability. Calculate the test statistic $T = \sum_{i=1}^{6} \frac{(n_i - n/6)^2}{n/6}$ which is known to have a Chi-square distribution with 6 degrees of freedom. Use S-PLUS to calculate the critical value for the Chi-square distribution at the 5% level. Can you reject the hypothesis that all numbers occur with the same frequency? Repeat this exercise with different numbers of rolls of the die.

Exercise 6.5

In this exercise, we will use simulation to examine a statistical hypothesis test. We want to test the null hypothesis that the mean of a Normally distributed sample is 0, with known variance 1. Let the alternative be that the mean is not equal to 0 (two-sided).

We know that if we set the significance level to α (alpha), the test will reject the null hypothesis in α percent of the times, although the data actually come from a Normal distribution with mean 0 and variance 1. Prove this by generating 1000 samples of size 100, calculating for each sample whether or not the null hypothesis is rejected, and summarizing how often the null hypothesis was rejected.

The other influence factor for a test is the power, the probability of sticking with the null hypothesis although the alternative is correct. Calculate the power of the test for the alternative point hypothesis that the mean is equal to 0.1.

Exercise 6.6

Suppose we had forgotten the formula for computing the area of a circle. All is not lost. We can easily use the computer to approximate a circle's area by simulation. For convenience, we want to determine the area of the unit circle with radius 1. We name this unknown constant Pi (or π).

We know that the area covered by a square with the corner points (-1,1), (1,1), (1,-1) and (-1,-1) equals 4, as the length of each side is 2. This square completely contains the unit circle. If we had a random point somewhere in the square, we could calculate the probability that the point is also inside the unit circle. A random point Z is determined by two coordinates, say X and Y, which have independent uniform distributions on the interval [-1,1].

The probability of Z being inside the circle is given by the ratio of the circle's area and the square's area or, more mathematically, letting P=probability:

P(Z is inside the circle) = (area of the circle)/(area of the rectangle) = Pi/4.

Knowing this, we can estimate (the unknown) Pi by multiplying the estimated probability for Z being inside the circle, by 4.

What's left is to generate a large number of random points Z, calculate the proportion of points inside the circle, and multiply this by 4. This is our estimate of Pi.

Write an S-PLUS program for this problem and estimate Pi with 100, 1000, and 10,000 random points. Try to work with complete vectors only and avoid loops. How far is the approximation away from the Pi we (fortunately) happen to know?

Hint: You need to know how to check if a point (x,y) is inside the circle. As the circle is defined by its origin and the radius, calculate how far the distance is from the circle's origin to the point. Use Pythagoras's rule to determine the distance between the middle of the circle and a point.

6.7 Solutions

Solution to Exercise 6.1

This exercise shows what different samples coming from the same distribution can look like. If you compare the graphs in the same column, the data are always exactly the same, but the parameter bin width for the histogram is changed. If you compare two graphs in the same row, the difference comes from the sample size (100 versus 1000), and, of course, from stochastic variation. The figure created is shown in Figure 6.15.

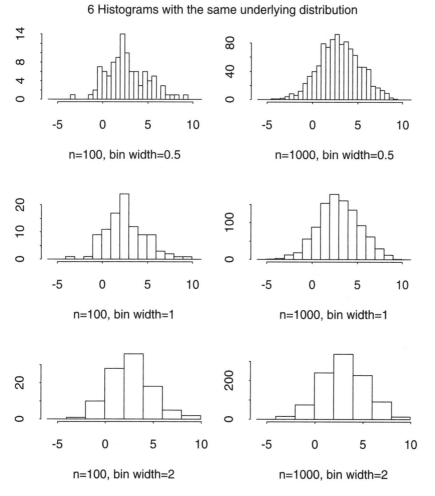

Figure 6.15. Different histogram displays of Normally distributed data.

We start by creating the two data sets.

```
> x <- rnorm (100, 3, sqrt (5))     # generate the data
> y <- rnorm (1000, 3, sqrt (5))
```

We determine the data range and divide the intervals into equally spaced subintervals ranging from the minimum to the maximum value, having different widths for the bins. Note that we need to calculate the *common* minimum and maximum over the two data sets to cover the whole range for both data sets. This generates three sequences of boundaries.

```
> min.xy <- min (x, y)              # determine the range for
> max.xy <- max (x, y)              #  the histogram bounds
> breaks1 <- seq (min.xy, max.xy, 0.5)
> breaks2 <- seq (min.xy, max.xy, 1)
> breaks3 <- seq (min.xy, max.xy, 2)
```

Now we are ready to actually do the histograms using the data we generated above. We change the character size and in particular, we set the histogram color as the background color, which is always the color number 0.

```
> par (mfrow=c(3,2), cex=0.6)       # the page layout
> hist (x,breaks=breaks1,xlab="n=100, bin width=0.5",col=0)
> hist (x,breaks=breaks2,xlab="n=100, bin width=1",  col=0)
> hist (x,breaks=breaks3,xlab="n=100, bin width=2",  col=0)
> hist (y,breaks=breaks1,xlab="n=1000,bin width=0.5",col=0)
> hist (y,breaks=breaks2,xlab="n=1000,bin width=1",  col=0)
> hist (y,breaks=breaks3,xlab="n=1000,bin width=2",  col=0)
```

Finally, we set a title on top of the layout figure using the `mtext` function. Experimenting with the position (the line) of the title shows that the line -1 is a good position to place the title. The side number 3 is the top of the figure, and setting `outer=T` puts the title on top of the overall figure instead of on top of the current (sub-)plot.

```
> title.string <-
+ "6 Histograms with the same underlying distribution"
> mtext (title.string, side=3, outer=T, line=-1)
```

Solution to Exercise 6.2

In this example, we will work with missing data and infinite values, to understand how they are treated and how you can actually handle them in data sets.

We begin by creating a vector by entering it directly from the keyboard.

```
> x <- c(1, 2, NA, 4, 1/0)
```

We could have entered `+Inf` or `-Inf` in the same way. Now let's try to calculate the mean of these data directly. This results in

```
> mean (x)
     NA
```

As we have learned in this section, we can specify a parameter for removing the missing data before doing the calculation:

```
> mean (x, na.rm=T)
     Inf
```

We see that S-PLUS removed the missing value, but not the infinite value. And, of course, the sum of data with an infinite value is infinite (it works vice versa with negative infinite data). To calculate the mean (and variance) over all *real* numbers, we need to remove the infinite value, too.

```
> y <- x[!is.na (x) & !is.inf (x)]
> mean (y)
     2.333333
> var (y)
     2.333333
```

Now let's divide x by itself:

```
> x/x
     1 1 NA 1 NA
```

Note that a missing value divided by anything else results in a missing value again, and infinity divided by infinity also results in a missing value.

Solution to Exercise 6.3

We do the Car data analysis interactively. The first thing we do is attach the data set `car.all`.

```
> attach (car.all)
```

To determine which variables are in the data set, the **names** function delivers the data set's column labels or, in other words, the variable names.

```
> names (car.all)
     "Length" "Wheel.base" "Width" "Height" "Front.Hd."
     "Rear.Hd" "Frt.Leg.Room" "Rear.Seating" "Frt.Shld"
     "RearShld" "Luggage" "Weight" "Tires" "Steering"
     "Turning" "Disp." "HP" "Trans1" "Gear.Ratio"
     "Eng.Rev" "Tank" "Model2" "Dist.n" "Tires2"
     "Pwr.Steer" ".empty." "Disp2" "HP.revs" "Trans2"
     "Gear2" "Eng.Rev2" "Price" "Country" "Reliability"
     "Mileage" "Type"
```

Since we have already attached the data set, we can access the variables just by their names. For example, entering HP returns the HP data for all the cars in the data set.

The names of the cars in the data set are returned by the names of the rows of the data matrix. Therefore, we enter

```
> dimnames (car.all)
```

to see the car names and the variable names, or just

```
> dimnames (car.all)[[1]]
```

to get only the first component of the list, the names of the cars. Now we save the names of the cars for later use.

```
> car.names <- dimnames(car.all)[[1]]
```

Next, we calculate the maximum travel distance for all cars.

```
> max.travel <- Mileage*Tank
```

To extract the categories and the car with the longest maximum travel distance for each category, we need to figure out what the categories are and use the by function to do the calculation by category. We first figure out that the variable's name for the category of car is Type. We can use the summary command to view the data structure.

```
> summary (Type)
  Compact  Large  Medium  Small  Sporty  Van  Na's
      19      7      26     22      21   10     6
```

We see that there are only seven cars classified as large in the data, six have no classification, and there are more than twenty cars in the groups Medium, Small, and Sporty.

Determining the cars with the highest fuel consumption (or the lowest mileage) in the different categories is done by using min in combination with by as follows.

```
> by (Mileage, Type, min, na.rm=T)
  Compact  Large  Medium  Small  Sporty  Van
      21     18      20     25      24   18
```

Note that we need to supply the parameter na.rm=T to the min function, as there are missing values in the data. The minimum will be missing if any of the data are missing.

In the same fashion, we obtain the maximum values.

```
> by (Mileage, Type, max, na.rm=T)
  Compact  Large  Medium  Small  Sporty  Van
      27     23      23     37      33   20
```

You might have produced slightly different output, because we used the command unclass(by(...)) to obtain this shortened output.

To analyze the four variables HP, Tank, Mileage, max.travel, we create a new data set containing only these data.

```
> car.subset <- data.frame (HP, Tank, Mileage, max.travel)
```
A first display of the data can be generated using `pairs`.
```
> pairs (car.subset)
```
The display of pairwise scatterplots is shown in Figure 6.16. We immediately

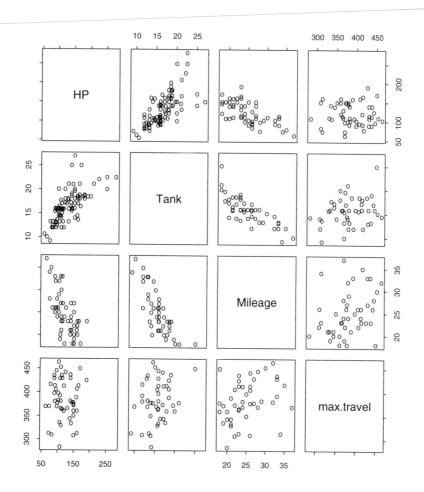

Figure 6.16. Some components of the Car data set (horse power, tank size, mileage, and maximum travel distance).

see strong relationships between many variables, like between HP and tank size or between tank size and mileage in a negative way: the smaller the mileage, the larger the tank, and vice versa.

Confirming this by numbers is a little harder to do because of the missing data contained in the Car data. We want to calculate the correlation between the variables using `cor`, or the variance/covariance matrix using `var`. Neither of the functions accept missing data, which is sensible. Therefore, we need to remove all car data with a missing value in one or more of the four variables. The function `is.na` is useful for checking if any data are missing, and if you remember that `TRUE` values are coded with 1 and `FALSE` values with 0, then it becomes clear what to do. We sum up the `TRUE/FALSE` values obtained from `is.na` and see if the sum is 0 (indicating that none of the data are missing) or greater than 0. The latter data are removed.

```
> problems <- apply (is.na (car.subset), 1, sum) != 0
> car.without.missings <- car.subset [!problems, ]
```

Now we can actually calculate the correlation matrix.

```
> cor (car.without.missings)
                    HP     Mileage        Tank   max.travel
       HP    1.0000001   -0.666153    0.676410   -0.0205656
  Mileage   -0.6661530    1.000000   -0.801337    0.3727046
     Tank    0.6764099   -0.801337    1.000000    0.2121667
max.travel  -0.0205656    0.372705    0.212167    1.0000000
```

The results of our first guess can be immediately confirmed, as the correlation between HP and tank size is 0.68. There is also a strong negative correlation between tank size and mileage, about -0.8.

Solution to Exercise 6.4

The dice test is done by generating a simulation of die rolls. There are several ways to simulate a die roll. A standard solution is to generate uniform random numbers on the interval from 0 to 6, cut off the digits behind the comma using `trunc`, and add 1 to obtain a sample of numbers from 1 to 6.

We show a different approach using the S-PLUS function `probsamp` to generate a sample from a set of values (sampling with replacement), and the function `table` will count the outcomes for us.

```
> n <- 6000
> rolls <- sample (1:6, n, replace=T)
> rolls.count <- table (rolls)
> rolls.count                       # the tabulated outcome
        1      2      3      4      5      6
      988   1028    988   1062    944    990
> test.stat <- sum ((rolls.count - n/6)^2 / (n/6))
> test.stat                         # the test statistic T
    8.152
```

```
> alpha <- c(0.9, 0.95, 0.99)
> qchisq (alpha, 6)                    # the critical values
     10.64464   12.59159   16.8119
```

As you can see, our test statistic is smaller than all three critical values of the Chi-square distribution, such that we cannot reject the hypothesis that we generated rolls with a fair set of dice.

In fact, the lowest alpha level at which the hypothesis could be rejected (the p-value) is calculated as

```
> 1 - pchisq(test.stat)
```

Solution to Exercise 6.5

You can simulate a statistical test's rejection percentage by generating data from the null hypothesis several times, and by calculating how often the test would reject the null hypothesis, although the data come from the distribution specified in the null hypothesis. In this exercise, we will only work on matrices and vectors, as this is an elegant and efficient way to program in S-PLUS.

We first generate 1000 samples of size 100 from the null hypothesis, being a Normal distribution with mean 0 and standard deviation 1. The samples are stored in a matrix, each row representing one sample.

```
> x <- matrix (rnorm (100000), 1000, 100)
```

Next, we calculate the mean and the standard deviation for each sample separately. Note that this is done without looping.

```
> m <- apply (x, 1, mean)
```

Finally, we need to check whether the test statistic - the mean of the sample - is outside the interval of the $(\alpha/2, 1-\alpha/2)$ quantiles of the test statistic's distribution. Knowing that the variance of a mean of identically independently distributed samples is reduced by the square root of the number of samples, we obtain that the test statistic comes (given the hypothesis) from a N(0, $1/\sqrt{100}) = \mathcal{N}(0, 0.1)$ distribution. As the Normal distribution is symmetric around the mean, we can also check if the sample mean's absolute value is larger than the $1 - \alpha/2$-quantile.

```
> z <- abs (m) > qnorm(0.975, 0, 0.1)
```

The result is a vector of TRUE/FALSE values, and summing them up yields the number of TRUE values, which in our case was 47. The mean value is the mean number of rejections over all 1000 samples, which should be around 0.05 (as we are testing on the 5% level).

```
> mean (z)
     0.047
```

You can also check what's in the z variable by tabulating the data.

```
> table (z)
    FALSE   TRUE
      953     47
```

To calculate the approximate power of the test when the data come from the alternative, or, in other words, the number of correct rejections of the null hypothesis, we generate samples from the alternative hypothesis and calculate how often the test rejects the null hypothesis.

```
> x <- matrix (rnorm (100000, 0.1, 1), 1000, 100)
> m <- apply (x, 1, mean)
> z <- abs(m) > qnorm(0.975,0,0.1) # number of rejections
> mean (z)
    0.185
```

We find that if the alternative hypothesis was that the mean is equal to 0.1, the Normal test rejects the wrong hypothesis correctly in only approximately 18.5% of the simulations. Of course the further the alternative is away from the null hypothesis, the better the test in terms of the power.

Another lesson from this is if the significance level (1 - alpha) were increased, the number of correct rejections of the null hypothesis would become even smaller, as the quantile of the Normal distribution on the righthand side of the comparison would become larger. In other words, the lower the type I error (rejecting the hypothesis although it is correct), the higher the type II error (not rejecting the hypothesis although it is incorrect).

A statistical test's power is typically illustrated by calculating many power values for different point alternatives and connecting the points by a curve to obtain the power curve. If you are interested, go ahead and calculate a power curve for this test. Guess what the curve looks like before actually calculating and plotting it.

Solution to Exercise 6.6

We show how to simulate 1000 two-dimensional random numbers in the square [-1,1] x [-1,1] and how to determine for each point if it is inside or outside the circle with radius 1 around the origin (0,0). Finally, we estimate π (Pi) from the proportion of points being inside the circle to the total number of points generated.

```
> n <- 1000
> x <- runif (n, -1, 1)
> y <- runif (n, -1, 1)
```

Determining if a point is inside the unit circle is equal to determining if the distance between the point and the origin of the circle (0,0) is less than 1, as we know that all points inside a circle are at maximum the radius away from the origin of the circle. Calculating the distance between two points in two dimensions is done by using Pythagoras' rule, and the distance between the points $x = (x_1, x_2)$ and $y = (y_1, y_2)$ is given as $\sqrt{(x_1 - y_1)^2 + (x_2 - y_2)^2}$. We omit the square root, as the inequality is untouched, because the radius on the righthand side is equal to 1.

```
> distance <- x^2 + y^2
> inside.circle <- sum (distance <= 1)
```

Now we estimate the unknown constant π according to the formula derived before.

```
> pi.estimate <- (4*inside.circle)/n
```

The constant pi is an S-PLUS constant, such that we can calculate how far we are away from the correct value of π (that we fortunately happen to know).

```
> away.from.pi <- pi - pi.estimate
```

Just to show that it is also possible to do the whole calculation on a single line, here is our one-liner:

```
> pi.est <- (length(runif(n,-1,1)^2+runif(n,-1,1)^2 <= 1)*4)/n
```

In the following table, we show what we obtained as estimates for different numbers of random values. Depending on the current sequence of random numbers, you might get different results. What you should note, nevertheless, is that the deviation from the true Pi gets smaller as n increases.

Table 6.9. Estimating Pi using a simulation technique. Results for different sample sizes

n	Estimate	Error
100	2.96	0.182
1000	3.088	0.0535
10000	3.1532	-0.0116
100,000	3.1366	0.00499

To visualize what we just simulated, plot the square and the circle and add the simulated points to it. The result might look like Figure 6.17. We draw the geometric figures rectangle and circle, and add the points to it. This might look like the following little piece of code.

```
> corners.x <- c(-1,-1, 1, 1, -1)
> corners.y <- c(-1, 1, 1,-1, -1)
> circle.seq <- seq (0, 2*pi, length=1000)
> plot(corners.x,corners.y,type="l",xlab="",ylab="",axes=F)
> axis(1)
> axis(2)
> lines (sin (circle.seq), cos(circle.seq))
> points (x, y)
> title ("Estimation of Pi using simulation")
```

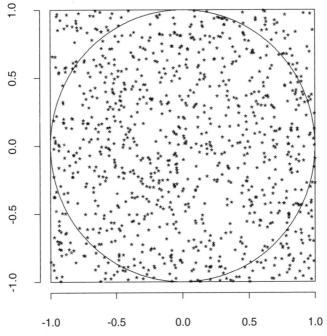

Figure 6.17. Estimating Pi using a simulation technique.

7. Statistical Modeling

We have now learned some elementary statistical techniques in S-PLUS and the basics of graphical data analysis. The next step is to see what S-PLUS has to offer in terms of modeling. Statistical modeling is one of the strongest S-PLUS features because of its unified approach, wide variety of model types, and excellent diagnostic capabilities. We start with an example of how to fit a simple linear regression model and corresponding diagnostics. The example is presented with a minimum of technical explanation, designed as a quick introduction. We then formally explain the unified approach to model syntax and structure, along with comments on several of the more popular types of statistical models.

7.1 Introductory Examples

We start with a simple linear regression model to show how to carry out the fit and what the output looks like.

7.1.1 Regression

For our example, we will use the Swiss fertility data set that is directly accessible from within S-PLUS by typing swiss.fertility and swiss.x, respectively. We chose this data set because it is easy to understand and work with. The data come from French-speaking provinces in Switzerland from about the year 1888, so no comparison to the present day should be made. Table 7.1 describes the variables.

We will restrict our discussion here to only one predictor variable, education, so we will not use the entire swiss.x matrix. The outcome of interest is the vector, swiss.fertility. Linear regression (simple or multiple) is done using the lm function. All you have to supply is the outcome variable followed by the predictor variable.

```
> education <- swiss.x[, "Education"]
> fit <- lm (swiss.fertility ~ education)
> summary (fit)
```

Table 7.1. Swiss fertility data set

Variable	Description
swiss.x	Agriculture: % of population with this occupation
	Examination: % of soldiers with high marks on
	army examination
	Education: % with education beyond primary school
	Catholic: % of population with this religion
	Infant Mortality: % of live births who live less than 1 year
swiss.fertility	Standardized fertility measure I[g]

```
Call: lm (formula = swiss.fertility ~ education)
Residuals
      Min    1Q Median    3Q    Max
   -17.04 -6.711 -1.011 9.526 19.69

Coefficients:
                Value Std. Error  t value Pr(>|t|)
  (Intercept) 79.6101    2.1041   37.8357   0.0000
    education -0.8624    0.1448   -5.9536   0.0000

Residual standard error: 9.446 on 45 degrees
   of freedom
Multiple R-Squared: 0.4406
F-statistic: 35.45 on 1 and 45 degrees of freedom,
   the p-value is 3.659e-007

Correlation of Coefficients:
            (Intercept)
education -0.7558
```

It is convenient to put the fit of the model into a variable because, as we will see later, model building and diagnostic routines use the information contained in it.

Notice that the first part of the output from the fit of the regression model contains basic summary information about the residuals. This is one simple way of checking the fit of the model. If the residuals are skewed in any way, then chances are that more formal diagnostic techniques will confirm that the model does not fit well.

The next section of the output contains information on the coefficients, i.e. the specific terms included in the model. Included are the estimates of the coefficients, standard errors, t-statistics, and p-values with which you can assess the fit of any of the specific terms in the model. An intercept is included in the model by default, but this may be removed. Note that in the fit of the model, both the intercept and education are statistically significant. In practical terms, this means that the education level is significantly related to the fertility level. Further, the coefficient of education is negative (-0.8624), implying that an increase in education is associated with a decrease in fertility. With simple linear regression, you can take the square root of the R-squared value (here 0.4406) to obtain a correlation of -0.6638. The `cor` function yields the same result.

```
> cor (education, swiss.fertility)
         -0.6637889
```

The `summary` function also gives the residual standard error, R-squared, and the overall F-test. The residual standard error is used in the calculation of many regression tests, the R-squared can be helpful in assessing the degree of association between outcome and explanatory variables, and the overall F-test is used for the entire model, as opposed to each separate factor.

7.1.2 Regression Diagnostics

At a first glance, the above regression model appears to fit the data quite well. However, as part of a proper and thorough regression analysis, you should check that the model assumptions are met. You can easily do this by examining the residuals that are contained in the object, `fit`. Four easy graphs to assess the model fit and check the assumptions include a scatterplot of the data, a histogram of the residuals, a scatterplot of the residuals, and a QQ-plot of the residuals. The commands for obtaining these graphs are shown below. The graphs appear in Figure 7.1.

Example 7.1. Regression Diagnostics

We first set up the graphics layout to contain two rows and two columns of plots. Our first plot is a simple scatterplot of the data with the fitted regression line superimposed.

```
> par (mfrow=c(2, 2))
> plot (education, swiss.fertility)
> abline (fit)
```

Our second plot is a histogram of the residuals to see if they are approximately Normally distributed.

```
> hist (resid (fit))
```

The third plot is of the predicted values against the residuals. To this we have put the reference line at y=0 and a `lowess` curve (a scatterplot smoother). This will tell us if the variance is a function of the covariate or some other problem with the fit of the model.

```
> plot (fit$fitted.values, resid (fit))
> lines (lowess (fit$fitted.values, resid (fit)))
> abline (h=0)
```

The last plot is a QQ-plot with the expected line to tell us whether or not the residuals follow a Normal distribution.

```
> qqnorm (resid (fit))
> qqline (resid (fit))
```

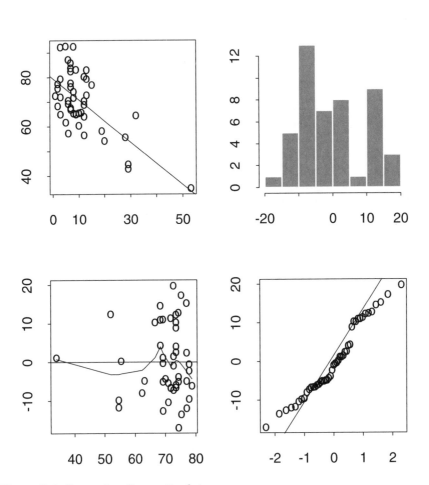

Figure 7.1. Regression diagnostic plots.

The scatterplot of the data shows that there is indeed an approximately linear relationship between the two variables, which is accentuated by the estimated regression line that has been superimposed. A histogram of the residuals would ideally look like the bell-shaped curve of the Normal distribution. The histogram here, with a little imagination, is not so far away from that ideal. The scatterplot of the residuals versus fitted values (values predicted from the regression equation) should show no discernable pattern. A lowess curve has been added to help detect whether or not hidden (or obvious) trends exist in the residuals. The flat curve indicates that perhaps the assumptions of the linear regression are met. A last check is to use a qqplot to assess whether the residuals follow an approximately Normal distribution. The superimposed straight line is what we would expect if the residuals followed a Normal distribution exactly. What we find is that our residuals, although not perfectly following a Normal distribution, are not that far away from the line.

7.2 Statistical Models

S-PLUS offers many possible ways of modeling data according to the assumptions made on the distribution and type of outcome (see Table 7.2). Simple linear regression is done with the lm function, as shown in the example in the previous two sections, and analysis of variance (ANOVA) with the aov function. ANOVA is actually a regression model, but the aov function makes it easier to define the classification variables used in these models.

Table 7.2. Selected statistical models in S-PLUS

Function	Type of Model
lm	Linear regression
aov	ANOVA
glm	Generalized linear models
gam	Generalized additive models
loess	Local regression smoother
tree	Classification and regression tree models
nls, ms	Nonlinear models
manova	Multivariate ANOVA
factanal	Factor analysis
princomp	Principal components analysis
crosstabs	Cross-classifications
survfit	Kaplan-Meier survival curves (formerly surv.fit)
coxph	Cox proportional hazards models (formerly coxreg)

The glm function is based on exponential family theory and can be used to do logistic, poisson, and other types of non-Gaussian "regression." Such regression models are called generalized linear models. The gam function extends the glm methodology by allowing nonparametric functions to be estimated from the data using smoothing operations. A local regression model is done with the loess function, which is a very convenient way of gaining an understanding of the relationships in the data. Classification and regression tree models can be run with the tree function, and nonlinear models with the `nls` and `ms` functions. Classical multivariate models, which include multivariate ANOVA, factor analysis, and principal components analysis, can be fit with the respective functions of `manova`, `factanal`, and `princomp`. Loglinear modeling can be done with the `crosstabs` function. And finally, for our short list, the most common survival analysis procedures can be done using the `survfit` and `coxph` functions.

7.3 Model Syntax

A particularly nice feature of S-PLUS is that it takes a unified approach to modeling, including the model syntax. All statistical models in S-PLUS specify the response and explanatory variables in the same manner, namely, Response ˜ Explanatory. The tilde (˜) can be read as "is modeled by" or simply as "equals." If we have the response variable, `NumCars`, and the explanatory variable, `Income`, and we want to model their relationship, we use the simple formula:

NumCars ˜ Income

We discuss the slightly more complicated matter of how to use the formulas in the following sections. For the moment, we concern ourselves only with the syntax of the model formula.

If you also have information on age and want to incorporate this into the model, simply add it with a plus sign (+) as in:

NumCars ˜ Income + Age

Interactions are usually of interest and are specified using the colon (:) as shown in the following example.

NumCars ˜ Income + Age + Income:Age

S-PLUS has several more advanced ways of specifying formulas for models. They are listed, along with those already discussed, in Table 7.3.

Table 7.3. Model syntax

Terms Included	Functionality
A + B	Main effects of both A and B
A:B	Interaction between A and B
A*B	Shorthand for: A + B + A:B
A %in% B	A nested in B
A/B	Shorthand for: A + B %in% A
A − B	All of A except what is in B
A ˆ m	All terms in A and m-order interactions

7.4 Regression

We now show, very briefly, how to use elementary regression functions. We include in the descriptions some commonly used modeling and model-building techniques. The nonparametric regression functions are, in particular, one of S-PLUS's major strengths, and you should consult the manual for detailed explanations.

There are, in principle, two types of models fit in S-PLUS, categorized by what is returned. Functions like linear regression return the *parameters of the fit*; that is, they return what is needed to determine the functional form. On the other hand, the nonparametric functions like `ksmooth` (kernel smoothing) return a list of points, x and y, where y contains the *smoothed data at points* x, as there is no explicit functional form for these types of smoothers.

For the first type of model, it is helpful to plot the data with

```
> plot (x, y)
```

and do the fit with, for example, `lm`,

```
> fit <- lm (y ~ x)
```

and then to add the fitted line with the command:

```
> abline (fit)
```

These commands were shown in Section 7.1.2.

In the second case, where the return argument consists of a list with x and y component (and others), you enter the following commands instead:

```
> plot (x, y)
> fit <- ksmooth (y ~ x)
> lines (fit)
```

7.4.1 Linear Regression and Modeling Techniques

As we already presented simple linear regression in Section 7.1.1, we use this section to review the unified modeling syntax presented in Section 7.3, and some of the model-building techniques available in S-PLUS. The examples in this section use the Swiss fertility data used earlier.

To begin, we simply replicate the model shown in Section 7.1.1. The model consisted of the response variable, `fertility`, and the explanatory variable, `education`.

```
> fit1 <- lm (swiss.fertility ~ education)
```

If the data had been in a data frame, we could have used the `data` option. A table summarizing the fit of the regression is obtained with the `summary` function as seen below:

```
> summary (fit1)

  Call: lm (formula = swiss.fertility ~ education)
  Residuals:
      Min    1Q Median    3Q    Max
   -17.04 -6.711 -1.011 9.526 19.69

  Coefficients:
                Value Std. Error  t value Pr(>|t|)
  (Intercept) 79.6101    2.1041   37.8357   0.0000
    education -0.8624    0.1448   -5.9536   0.0000

  Residual standard error: 9.446 on 45 degrees
    of freedom
  Multiple R-Squared: 0.4406
  F-statistic: 35.45 on 1 and 45 degrees of freedom,
    the p-value is 3.659e-007

  Correlation of Coefficients:
              (Intercept)
  education -0.7558
```

Note that the output here is exactly the same as shown in the example in Section 7.1.1 except that now we will build on this simple model.

With the unified approach to modeling, we can very easily make use of many sophisticated techniques. For example, suppose we decided that education was not adequate to describe fertility and we wanted to add more terms to the model. We want to update the previous model, and this is done, appropriately, with the **update** function as follows:

```
> catholic <- swiss.x[, "Catholic"]
> fit2 <- update (fit1, . ~ . + catholic)
> summary (fit2)

  Call: lm (formula = swiss.fertility ~ education +
      catholic)
  Residuals:
      Min    1Q Median    3Q    Max
   -15.04 -6.576 -1.426 6.126 14.32

  Coefficients:
                Value Std. Error  t value Pr(>|t|)
  (Intercept) 74.2319    2.3518   31.5636   0.0000
    education -0.7882    0.1293   -6.0968   0.0000
    catholic  0.1109    0.0298    3.7224   0.0006
```

```
Residual standard error: 8.331 on 44 degrees
  of freedom
Multiple R-Squared: 0.5746
F-statistic: 29.71 on 2 and 44 degrees of freedom,
  the p-value is 6.822e-009

Correlation of Coefficients:
            (Intercept) education
education -0.6838
  catholic -0.6144        0.1540
```

Note that in the updated model, fit2, both education and catholic are significantly related to fertility. Our choice of adding the variable, catholic, was more or less random, but we could have used built-in functions to make the choice a little less whimsical. The add1 function takes a list of explanatory variables and adds them one at a time to the current model. The resulting output shows statistics for each explanatory variable, had each been included together with the base model.

```
> agric <- swiss.x[, "Agriculture"]
> exam <- swiss.x[, "Examination"]
> inf.mort <- swiss.x[, "Infant Mortality"]
> add1 (fit1, . ~ . + agric + exam + catholic + inf.mort)

  Single term additions

  Model:
  swiss.fertility ~ education
            Df Sum of Sq      RSS        Cp
    <none>                4015.236 4372.145
     agric  1    61.9657 3953.270 4488.635
      exam  1   465.6258 3549.610 4084.975
  catholic  1   961.6083 3053.627 3588.992
  inf.mort  1   891.2462 3123.989 3659.354
```

Applied to our data, we see that with the largest sums of squares and the lowest Cp value, the variable catholic actually does provide the best predictive power when one variable is added to a model that includes education.

On the other hand, we could have chosen to use stepwise regression with the step function, or chosen to add a lot of variables and remove them using the drop1 (opposite of add1) function. The choice between using add1, drop1, or step is left entirely up to you. In general, they will yield the same model, but often in practice, they do not. So, have fun.

S-PLUS offers a host of residual checking techniques. Feel free to use the simple guidelines shown in Section 7.1.2, but you can do some more advanced work by simply plotting the fit from the regression.

```
> plot (fit1)

Make a plot selection (or 0 to exit):

1: plot: All
2: plot: Residuals vs Fitted Values
3: plot: Sqrt of abs(Residuals) vs Fitted Values
4: plot: Response vs Fitted Values
5: plot: Normal QQplot of Residuals
6: plot: r-f spread plot
7: plot: Cook's Distances
```

This command gives you the above menu from which you can choose whichever model-fitting diagnostic you prefer. A hint for requesting all of the plots is to use par (mfrow=c(3, 2) before the plot command. This sets up the graphics window to make six graphs.

S-PLUS can calculate the predicted values, standard errors, and confidence intervals using the predict, se.fit, and pointwise functions.

A variable's polynomial effects are obtained with the poly function. To include linear, quadratic, and cubic terms of X in a model, use poly (X, 3), but beware that the resulting parameter estimates are in terms of orthogonal polynomial contrasts and are not on the same scale as X. To see the parameter estimates on the original scale, use

```
> poly.transform (poly (X, 3), coef (fit))
```

Indicator Variables. S-PLUS can automatically generate indicator variables using the I function. Entering the term I(age > 70) into your model is a way of looking at the effect of "young" vs. "old" rather than the effect of age on a continuous scale.

Factors. Factors are related to the issue of indicator variables. You may often have variables like political affiliation or treatment that can assume one of several values. Giving numeric codes to these types of variables and entering them into your model as continuous variables makes little sense. To see the problem here, consider the following example. Suppose a questionnaire had been done collecting information on political affiliation, where 1=Democratic, 2=Republican, 3=Independent, and 4=Other. If we entered the continuous variable political affiliation (values 1 to 4), we would be making assumptions like Democratic is less than Republican, the difference between Democratic and Republican is the same as the difference between Republican and Independent, and "Other" is the highest. These assumptions, however, have no meaning and are of little interest.

By defining these categorical variables as factors, it is possible to view the effect of each level of the variable separately, or combined in certain ways using contrasts. Factor is a special data type in S-PLUS and is defined with the factor function.

The data set we have used so far doesn't contain any categorical variables, but we can easily create one using the cut function as follows.

```
> educ.lev <- factor (cut (education, breaks =
+ c(0, 5, 10, 100), labels = c("Low", "Average", "High")))
> is.factor (educ.lev)
      T
```

The optional breaks specifies at which values of education to define the breaks for the levels, and labels defines the names for the new categories. We have created a new variable with three levels of education; low is from 0 to 5, average is from 5 to 10, and high is from 10 to 100 (i.e., over 10). The function is.factor can be used to determine if your variable is really a factor.

Try fitting a regression model with educ.lev and compare the results to the model with education. What do you notice? Remember that education is a continuous variable but that educ.lev is a categorical variable with three levels. To make things even more complicated, you could convert educ.lev into a numeric variable with the three values of 1, 2, and 3 using both the unclass and the as.vector functions. How does this compare to the other two?

7.4.2 ANOVA

The analysis of variance (ANOVA) is a statistical method of analyzing data that assumes a continuous outcome variable and one or more classification variables. You might wonder how this differs from using a linear regression model with a factor variable. It doesn't. Theoretically speaking, an ANOVA model is a regression model. The difference comes more with the presentation of the results, as we see in the following example.

We use the Swiss fertility data, but to do so, we must convert the education variable into one with levels. We have already done this in the previous section by creating the variable educ.lev, which contains the three levels of low, average, and high levels of education.

The aov function is used exactly in the same fashion as the lm function to fit the desired model.

```
> fit.a <- aov (swiss.fertility ~ educ.lev)
```

The summary function is used to display the results, or you can use the version tailored to this class of model, the summary.aov function. In this case, the two versions of displaying the results are the same: the classical ANOVA table with degrees of freedom, sums of squares, mean squares, F-values, and corresponding p-values.

```
> summary (fit.a)

            Df Sum of Sq  Mean Sq  F Value      Pr(F)
educ.lev     2  1350.891 675.4456 5.100272 0.01018307
Residuals   44  5827.064 132.4333

> summary.aov (fit.a)

            Df Sum of Sq  Mean Sq  F Value      Pr(F)
educ.lev     2  1350.891 675.4456 5.100272 0.01018307
Residuals   44  5827.064 132.4333
```

We see from the large F-value and very low p-value that education is related to fertility, but we already knew this from the regression model. Is there any information we can obtain from the ANOVA model that we haven't already seen with the regression model? Of course!

With a factor variable, you can split the degrees of freedom into contrasts of interest. Take the `educ.lev` variable from above which has three levels and hence two degrees of freedom. We could use one degree of freedom for a linear test and one for a quadratic test. This breakdown of the degrees of freedom leads to results that are comparable to using polynomial regression mentioned in Section 7.4.1. The difference is that with contrasts, you aren't restricted to polynomial transforms, but we show it here for simplicity. To obtain a polynomial breakdown of the degrees of freedom, use the `split` option together with lists, as below.

```
> summary (fit.a, split=list (educ.lev=list (L=1, Q=2)))

               Df Sum of Sq  Mean Sq  F Value     Pr(F)
educ.lev        2  1350.891  675.446  5.10027 0.0101831
   educ.lev: L  1     3.289    3.289  0.02483 0.8755007
   educ.lev: Q  1  1347.602 1347.602 10.17571 0.0026237
Residuals      44  5827.064  132.433
```

The linear and quadratic effects of education on fertility are displayed and tested, showing that it is not actually the linear component of the relationship between education and fertility that is statistically significant, but the quadratic one.

The main objective with many ANOVA models is merely to test for certain effects. Often, however, the actual coefficients of the terms are necessary. These are obtained using the `coef` function as shown in the example below.

```
> coef (fit.a)

    (Intercept) educ.lev1 educ.lev2
        70.70725      -1.135 -3.797745
```

The choice between the `lm` and `aov` functions is often a matter of preference. If you are used to using one of the two, you naturally continue to do so.

7.4.3 Logistic Regression

So far we have examined models for fitting data with a continuous outcome. Often though, the outcome is binary in nature representing success/failure, 0/1, or presence/absence. In such cases, a logistic regression model is used to fit the data. Such models are fit in S-PLUS using the `glm` (or `gam`) function. This function is a very general procedure, as it can be used to fit models in the exponential family including the normal, binomial, poisson, and gamma distributions.

To explore this type of regression, we will use the built-in data frame, kyphosis, which contains a suitable variable with a binary outcome. The data frame contains the four variables described below in Table 7.4.

Table 7.4. Kyphosis data frame

Variable	Description
Kyphosis	Presence/absence of the postoperative spinal deformity, kyphosis
Age	Age of child in months
Number	Number of vertebrae involved in surgery
Start	Beginning of the range of vertebrae involved

We want to see what relationship exists, if any, between the age, number, and start, and the subsequent presence of kyphosis. For example, are younger children more susceptible to kyphosis or more resilient?

We use the `glm` function to fit our logistic model. All we need to specify is the model equation, the exponential family type, and the data frame as follows.

```
> fit.kyph <- glm (Kyphosis ~ Age+Number+Start,
+ family=binomial, data=kyphosis)
```

The results from the logistic model are viewed in the same manner as before using the `summary` function, although the printed output is slightly different.

```
> summary (fit.kyph)

Call: glm (formula = Kyphosis ~ Age + Number + Start,
family = binomial, data = kyphosis)
Deviance Residuals:
      Min         1Q       Median        3Q       Max
  -2.312363 -0.5484308 -0.3631876 -0.1658653 2.16133

Coefficients:
                  Value Std. Error    t value
(Intercept) -2.03693225 1.44918287  -1.405573
        Age  0.01093048 0.00644419   1.696175
     Number  0.41060098 0.22478659   1.826626
      Start -0.20651000 0.06768504  -3.051043

(Dispersion Parameter for Binomial family taken to be 1)

    Null Deviance: 83.23447 on 80 degrees of freedom

    Residual Deviance: 61.37993 on 77 degrees of freedom

    Number of Fisher Scoring Iterations: 5

    Correlation of Coefficients:
          (Intercept)        Age     Number
    Age -0.4633715
 Number -0.8480574    0.2321004
  Start -0.3784028   -0.2849547  0.1107516
```

The function call is repeated at the top, followed by some summary information on the residuals, the model coefficients, general fitting information,
and the correlations between the model coefficients. The t-statistics given are
partial t-tests, meaning that all other factors in the model have been adjusted
for. Notice that p-values are not provided with the t-statistics, but this is an
easy calculation to do in S-PLUS. You learned how to do it in the previous
chapter.

The t-statistics indicate that age and the number of vertebrae do not
seem to be related to kyphosis, but that the starting vertebra of the surgery
is related to it. We know that the degrees of freedom to be used for the t-
tests are the same as for the residual deviance (77). In this case the number
of degrees of freedom is large, so the cutoff for the t-distribution is similar to
that of the Normal, i.e. $\pm$ 1.96 for the usual significance level of 0.05. Only
the t-statistic for the starting vertebra of the surgery has a value that is more
extreme than -1.96, implying that it is associated with kyphosis. To be a little

more precise and a lot more helpful, we have calculated both the cutoff level of this distribution and the p-value.

```
> qt (.975, 77)
        1.991254
> qt (.025, 77)
        -1.991254
> 2*pt (-3.05, 77)
        0.003137395
> 2*(1 - pt (abs (-3.05), 77))
        0.003137395
```

Using an overall level of significance of 0.05 (each tail gets 0.025), we have calculated both the upper and lower cutoffs for the t-distribution with 77 degrees of freedom, and also the p-value using two slightly different methods.

The modeling techniques covered in the previous sections can also be applied to logistic models. You may wish to try some of them out now, or simply wait until the exercises.

7.4.4 Survival Data Analysis

What do we mean by "survival data"? Quite loosely, we refer to data that record the time until the occurrence of an event. In cancer trials, for example, you measure the time (in days, months, etc.) until the patient's death. In this case, the event is death. The problem is that quite often the patients either withdraw from the trial, or are still alive at the end of it (rather good from the patients' point of view). The problem raised is that the actual time of the event is not observed for these patients, so their survival time is "censored." Special statistical techniques have been devised to analyze this type of data, including the famous Kaplan-Meier curves. Recently there has been a lot of statistical research in this area on the topic of martingales and martingale residuals. S-PLUS is one of the few software packages to include this functionality. This book is not, however, a statistics book, so we will stick to the very basics and it will be up to you to carry on with martingales should you so desire.

With survival data analysis, the outcome "variable" consists of two pieces, the time to event and an indicator of whether the time refers to an event or is a censored time. It is then typical to have some sort of group variable such as a treatment indicator. One such data set included with S-PLUS is the Leukemia data set, which contains three variables as described in Table 7.5. The data come from a trial to determine the efficacy of maintenance chemotherapy for acute myelogenous leukemia.

The survfit function can be used to fit a Kaplan-Meier survival curve, or other types of survival curves according to the options chosen. You must use the Surv option to define the composite outcome variable, followed by the explanatory variables and the name of the data frame.

Table 7.5. Leukemia data frame

Variable	Description
time	Time to remission after chemotherapy (in weeks)
status	Indicator of status: 1=remission time; 0=censored time
group	Treatment group: maintained or nonmaintained

```
> leuk.fit <- survfit (Surv (time, status) ~ group,
+ leukemia)
> summary (leuk.fit)

    Call: survfit (formula = Surv (time, status) ~ group,
          data = leukemia)
```

group=Maintained

time	n.risk	n.event	survival	std.err	lower 95% CI	upper 95% CI
9	11	1	0.909	0.0867	0.7541	1.000
13	10	1	0.818	0.1163	0.6192	1.000
18	8	1	0.716	0.1397	0.4884	1.000
23	7	1	0.614	0.1526	0.3769	0.999
31	5	1	0.491	0.1642	0.2549	0.946
34	4	1	0.368	0.1627	0.1549	0.875
48	2	1	0.184	0.1535	0.0359	0.944

group=Nonmaintained

time	n.risk	n.event	survival	std.err	lower 95% CI	upper 95% CI
5	12	2	0.8333	0.1076	0.6470	1.000
8	10	2	0.6667	0.1361	0.4468	0.995
12	8	1	0.5833	0.1423	0.3616	0.941
23	6	1	0.4861	0.1481	0.2675	0.883
27	5	1	0.3889	0.1470	0.1854	0.816
30	4	1	0.2917	0.1387	0.1148	0.741
33	3	1	0.1944	0.1219	0.0569	0.664
43	2	1	0.0972	0.0919	0.0153	0.620
45	1	1	0.0000	NA	NA	NA

The summary function applied to this class of data provides a radically different output than those from the previous regression models. Here you get the estimated survival probability at many time points, along with standard deviations and confidence intervals for each of the two treatment groups. The estimated survival for the nonmaintained group at 30 weeks is 0.29, whereas it is 0.49 at 31 weeks for the maintained group.

Printing the fit of the survival analysis directly yields certain summary statistics for each of the two groups, such as the estimated mean and median survival times.

```
> leuk.fit

    Call: survfit (formula = Surv (time, status) ~ group,
            data = leukemia)
                                                0.95 0.95
                  n events mean se(mean) median  LCL  UCL
    group=       11      7 52.6    19.83     31   18   NA
    Maintained
    group=       12     11 22.7     4.18     23    8   NA
    Nonmaintained
```

With this output we see that the median survival time for the maintained group is 31 weeks but is only 23 weeks for the nonmaintained group. The next obvious question is whether or not the difference in survival between these two groups is statistically significant. Using the `survdiff` function, we can conduct a statistical test for the difference between two survival distributions, as below.

```
> survdiff (Surv (time, status) ~ group, leukemia)

                         N Observed Expected (O-E)^2/E
    group=Maintained    11        7    10.69      1.27
    group=Nonmaintained 12       11     7.31      1.86

    Chisq= 3.4  on 1 degrees of freedom, p= 0.0653
```

This is only the beginning of S-PLUS's survival analysis capabilities. Try `help (survival)` to find out more options with these models and plotting routines, especially **coxph** for Cox Proportional Hazards regression models.

7.5 Exercises

Exercise 7.1

In the regression model presented in this chapter, we assumed that `education` was a continuous variable. Often, however, we encounter variables that have discrete levels and must perform a regression model with them. Create four `education` groupings of roughly equal size, and an indicator variable for each grouping. The first indicator variable should be equal to one for each observation in the first grouping and zero for all other observations. The other indicator variables are similarly defined. Perform a multiple regression with these new indicator variables in place of the continuous version of `education`. What is the mean education level per grouping? Does the mean fertility differ between the groupings? What additional conclusions can you make from this analysis over the one presented in the chapter?

Hint: Use the `quantile`, `cut`, and `cbind` functions.

Exercise 7.2

Consider the Geyser data set described in Section 3.6. You can always look at the help system to refresh your memory.

The problem is to approximately determine the duration of the next eruption based on the waiting time for tourists who want to see this event. Plot the data set and add a title and axis labels. Add the simple least squares regression line to the plot.

You should see that the data are separated into two clouds. Fit a separate linear regression for each of the two subsets and add the two regression lines to the plot. Join the two separate fits so that they form a single line. The first regression line should go up to x=70 and the second for values of x greater than or equal to 70. Finally, add a legend to illustrate the different fits.

Hint: Use the `par ("usr")` function.

7.6 Solutions

Solution to Exercise 7.1

To make our lives easier, we first extract the vectors of interest so that we don't have to deal with matrices.

```
> education <- swiss.x[, "Education"]
> fertility <- swiss.fertility
```

Next we extract the quantiles we need to define our groupings. The groups will be defined as (1) minimum - 25th quantile, (2) 25th - 50th quantile, (3) 50th - 75th quantile, and (4) 75th quantile - maximum. Since the quantiles are based on the data (empirical), they might not have exactly the same number in each group, but it should be very close. The actual quantiles are shown below and will act as the cutoff points for the groupings.

```
> quants.educ <- quantile (education,
+ c(0, .25, .50, .75, 1))
    0%  25%  50%  75%  100%
     1    6    8   12    53
```

We will use the cut function to use the cut-points defined by the quantiles to create a variable that contains a 1 for the first group, a 2 for the second group, and so on. Note that we have also used the include.lowest option to include the minimum value. Groups are normally defined starting with any value greater than the lower cut-point and up to and including the upper cut-point. This method's flaw is that the minimum value gets left out. By including it, the lower category now has a slightly different definition than the others, but it should have little effect. On the other hand, if we simply ignored the minimum value, we wouldn't be using all the data.

```
> cuts.educ <- cut (education, quants.educ,
+ include.lowest=T)
```

The tapply function must be used here (instead of apply) because cuts.educ has been put into a list. We see that the first group has three more values in it than the other groups, but that the groups are roughly balanced. We also see that the mean education level increases from 3.79 in the first group to 24 in the fourth group.

```
> tapply (education, cuts.educ, length)
  1+ thru 6   6+ thru 8   8+ thru 12   12+ thru 53
         14          11           11            11
> tapply (education, cuts.educ, mean)
  1+ thru 6   6+ thru 8   8+ thru 12   12+ thru 53
   3.785714    7.363636     10.72727            24
```

To make our indicator variables, we first make four vectors and fill them with zeros. We then put ones into the vectors in the appropriate elements.

```
> educ1 <- educ2 <- educ3 <-
+ educ4 <- rep (0, length (education))
> educ1[cuts.educ==1] <- 1
> educ2[cuts.educ==2] <- 1
> educ3[cuts.educ==3] <- 1
> educ4[cuts.educ==4] <- 1
```

We use the cbind function to put our indicator variables into a matrix to be
used in the regression.

```
> newx <- cbind (educ1, educ2, educ3, educ4)
> summary (lm (fertility ~ newx))
```

We have run into a problem with our regression model. Fortunately, this is a
typical problem that is easily fixed. You have to realize that, unless specified,
linear regression includes an intercept term. The problem is that, numerically
speaking, the columns of our design matrix are not linearly independent,
which results in a singular matrix (intercept = educ1 + educ2 + educ3
+ educ4).

```
    Error in lm.fit.qr(x, y): computed fit is
    singular, rank 4
    Dumped
```

As mentioned above, we can solve this problem of overparameterization by
simply removing one of the indicator variables. The choice of which one to
remove won't change the results, only the interpretation.

```
> newx <- cbind (educ1, educ2, educ3)
> summary (lm (fertility ~ newx))

    Call: lm(formula = fertility ~ newx)
    Residuals:
          Min       1Q   Median       3Q      Max
       -25.64   -6.186   -2.236    6.382    22.26

    Coefficients:
                     Value   Std. Error   t value   Pr(>|t|)
    (Intercept)    60.6364       3.3783   17.9486     0.0000
      newxeduc1    13.6994       4.5145    3.0345     0.0041
      newxeduc2    16.0818       4.7777    3.3660     0.0016
      newxeduc3     7.1000       4.7777    1.4861     0.1446

    Residual standard error: 11.2 on 43
    degrees of freedom
    Multiple R-Squared: 0.2479
    F-statistic: 4.725 on 3 and 43 degrees of freedom,
    the p-value is 0.006164
```

```
Correlation of Coefficients:
               (Intercept)   newxeduc1    newxeduc2
   newxeduc1     -0.7483
   newxeduc2     -0.7071       0.5292
   newxeduc3     -0.7071       0.5292       0.5000
```

There is no indicator for education level 4, but we can estimate the fertility rate for that group based on the model above. The estimated coefficients above are for the intercept and the remaining three education levels. Since we left education level 4 out of the model, we have set its effect, not its value, to zero. To get the estimated fertility levels, take the coefficient from the intercept and add the education level coefficients for estimates of 60.6+13.7=74.3, 60.6+16.1=76.7, 60.6+7.1=67.7, and 60.6+0.0=60.6 respectively. The overall test of whether or not education is related to fertility rate is given by the F-test with 3 and 43 degrees of freedom. Note that the p-value for this test is well below the standard cutoff of 0.05, indicating that there is a significant relationship between the two measures.

By construction, the mean education level increases in the four groupings, but the fertility decreases. Notice that there is little change in mean fertility level between the first two groups. This is reflected in the fact that the coefficients are nearly the same for the variables neweduc1 and educ2. This indicates that the true relationship between education and fertility is not exactly linear, but is well approximated by a linear model.

Solution to Exercise 7.2

The following code stores the data from the Geyser data set into a dependent variable y and an independent variable x. This is done merely for convenience.

```
> x <- geyser$waiting
> y <- geyser$duration
```

We now make a simple plot of the data specifying the axis labels and title, as well as restricting the x-axis to go from the value of 20 to the maximum value of duration.

```
> plot (x, y, xlim=c(20, max (x))
+ main="Fits for the Geyser Data Set",
+ xlab="Waiting Time" ,ylab="Eruption Length")
```

We then calculate the least squares fit and store the fit into the variable lsf1 for later use. By later use, we mean that we use it to draw the fitted regression line to our plot, which is done with the abline function.

```
> lsf1 <- lm (y ~ x)
> abline (lsf1, col=4, lty=4)
```

We have noticed that the data separates into two groups and we want to explore the relationship between `duration` and `waiting` times in the two groups. A convenient place to split the data is at x=70. We do separate regressions on the two data subsets by using the `lm` function's `subset` option. The fit for x less than 70 is saved in `lsf2` and for x greater than or equal to 70 in `lsf3`. Both lines are added to the plot using the `lines` function, but notice that we are using a different line type each time so we will be able to distinguish them from one another.

```
> lsf2 <- lm (y ~ x, subset=x<70)
> lsf3 <- lm (y ~ x, subset=x>=70)
> abline (lsf2, col=5, lty=5)
> abline (lsf3, col=6, lty=6)

> a1 <- coefficients (lsf2)[1]
> b1 <- coefficients (lsf2)[2]
> a2 <- coefficients (lsf3)[1]
> b2 <- coefficients (lsf3)[2]
```

Next we want to construct a regression that will meet at the point x=70. There are many ways of doing this including mathematically creating a "knot" at x=70. We will take an easy way out by using information we already have. Essentially all we will do is join the subset regression lines by using the y-intercepts and slopes from our two subset regressions and the boundaries of the current graph.

To get our line to go to the edge of the graph, we have to know the value of the graph boundaries, not just the minimums and maximums of the x- and y-values. The `par ("usr")` function (from the hint) does just that. With this information, we can start at the point x=70, and work to the left and right. We can use the y-intercept and slope from `lsf2` and calculate the estimated y-values at both the left boundary of the graph (`par("usr")[1]`) and at x=70. Connecting these two points will give us the first part of our joined regression line. Doing the same for the point x=70 and the x-value at the right of the regression line (`par("usr")[2]`) gives us the second part of the joined regression line. The `lines` function draws a line through the four points.

```
> xpoints <- c(par ("usr")[1], 70, 70, par ("usr")[2])
> ypoints <- c(a1+b1*xpoints[1], a1+b1*70, a2+b2*70,
+ a2+b2*xpoints[4])
> lines (xpoints, ypoints, lwd=3)
```

For clarity, we have added a legend to the graph so we know which regression line is which.

```
> descr <- c("Overall LS Fit", "Group 1 Fit",
+ "Group 2 Fit", "Combined Fit")
> legend (c(20, 65), c(0.7, 2.7), descr, col=c(4:6,1),
```

```
+ lty=c(4, 5, 6, 1), lwd=c(1, 1, 1, 3)))
```

Note that the joined regression line has a jump at x=70. This is because we have estimated a y-value at x=70 from both regressions. The `lines` function simply connects the points it receives.

Finally, Figure 7.2 is obtained.

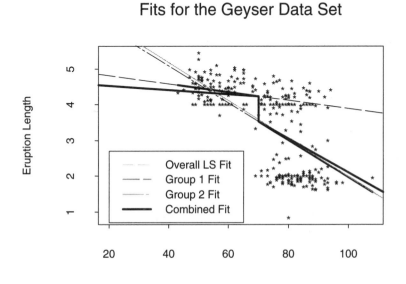

Figure 7.2. Different linear regression lines for the Geyser data set.

8. Programming

*"Writing program code is a good way
of debugging your thinking."*

Bill Venables on S-News
February 1997

Now that you have learned the elementary commands in S-PLUS and many ways of applying them, it is time to discover its advanced functionalities. This chapter introduces loops, deals more intensively with writing functions, covers debugging matters, and gives a short introduction to what object-oriented programming means in the S-PLUS environment.

8.1 Iteration

An iteration is, in principle, a loop or repeatedly executed instruction cycle, with only a few changes in each cycle. In programming languages that are not matrix or array-oriented, like Pascal, C, or Fortran, even a simple matrix multiplication needs three nested loops (over rows, columns, and the indices). Since S-PLUS is matrix-oriented, these operations are much more efficient and easy to formulate in mathematical terms. This means they are faster than loops and the code is much easier to read and write. Pay attention to the following guideline, which is elementary but needs to be kept in mind when writing S-PLUS code.

Note Whenever possible, try to avoid loops. 99% of the time, an operation on matrices is much more elegant as well as much faster. Try to use vectorized statements, or functions like `apply`, `lapply`, and `tapply`. ◁

Here is an easy example, calculated on an IBM compatible PC 486/50. If we define a vector x with 50,000 elements, and assign the square of x to y, without using a loop, the following two lines take about 2 seconds to execute.

```
> x <- 1:50000
> y <- x^2
```

On the other hand, using an explicit loop statement (**for**) over the values of x and y, respectively, results in the code

```
> x <- y <- 1:50000                    # initialization
> for (i in 1:50000) y[i] <- x[i]^2
```

and does exactly the same as the two lines above. You can see that the first expressions are easier to read, and on top of that, the second little program needs about 2 minutes to execute, or about 60 times longer than the first program.

We will treat the different forms of loops, **for**, **while**, and **repeat**, in more detail in the following section.

8.1.1 The **for** Loop

A loop construction is easy to understand by looking at a concrete example. The general syntax for a **for** loop is

```
> for (i in ivalues) { S commands }
```

and an easy example is

```
> for (i in 1:10) { print (i) }
```

If there is only a single statement after the **for** command, the brackets { } can be omitted, but for clear programming style, we recommend using the brackets.

The object **ivalues** can be of any type; a vector, a variable, a matrix, or anything else. S-PLUS walks through all the elements of the argument specified, replacing the loop variable (here i) by the elements successively. The looping variable **ivalues** can also be of any mode, like numeric, character, boolean, or a mixture of these.

Here are a few examples.

Example 8.1. A numeric looping variable

```
> for (i in c(3, 2, 9, 6)) { print (i^2)}
      9 4 81 36
```


Example 8.2. A character looping variable

```
> transport.media <- c("car", "bus", "train", "bike")
> for (vehicle in transport.media) {print (vehicle)}
        "car"
        "bus"
        "train"
        "bike"
```

| Note | There is no explicit return argument in loops. Thus, you must always use the **print** function to print values within a loop. Using a statement like

```
> for (x in 1:4) { x }
```

to print out the current value of x does not result in printouts of the values x takes. Only the very last expression of the loop is given back to the calling routine. So use always

```
> for (x in 1:4) { print (x) }
```

to print out something inside loops.   ◁

8.1.2 The while Loop

In contrast to the **for** loop, the **while** statement is used if the total number of iteration cycles is not known before the loop starts. For example, if you want to iterate until a certain criterion is fulfilled, the **while** loop is the appropriate statement to use.

Example 8.3. A simple while loop
 Calculate the sum over 1,2,3,..., until the sum is larger than 1000.

```
> n <- 0                         # the iteration counter
> sum.so.far <- 0                # store the added values
> while (sum.so.far <= 1000)     # the while loop
+ {
+   n <- n + 1
+   sum.so.far <- sum.so.far + n
+ }
```

The summed values are stored in **sum.so.far**, and the variable n is increased by 1 in each loop. Enter the short program yourself and let it run. What is the result? How many times did S-PLUS walk through the loop? You can also watch the sum converging to the result if you create an empty vector and store the sum in each loop in a vector element. A graph of the vector would then show the convergence.

There is a pitfall hidden here. You might think you can name the sum over the past values simply **sum**. As you know, there is an S-PLUS internal function of that name, such that we would have a function and a variable of the same name. Refer to page 24, where this situation is discussed further.

8.1.3 The repeat Loop

The third alternative for looping over expressions is to use the **repeat** statement. The **repeat** function repeats a set of expressions until a stop criterion

is fulfilled. In S-PLUS, the keyword **break** is used to tell the surrounding
repeat environment that it is time to stop.

Example 8.4. A simple repeat loop
We rewrite the code used above to demonstrate the **while** loop.

```
> n <- 0                            # the iteration counter
> sum.so.far <- 0                   # store the added values
> repeat
+ {
+   n <- n + 1
+   sum.so.far <- sum.so.far + n
+   if (sum.so.far > 1000) break
+ }
```

Another keyword inside **repeat** loops is **next**. If S-PLUS comes across the
next statement in a **repeat** loop, it stops executing the loop statements and
re-enters the loop from the top.

You might have realized that you could very easily put S-PLUS into an
infinite loop by simply forgetting to put in the **break** statement.

> [Note] The main difference between a **while** loop and a **repeat** loop is that
it is possible not to enter a **while** loop at all, if the condition is not fulfilled
at the beginning, whereas a **repeat** loop is always entered at least once.

Experienced programmers might know that every **repeat** construction
can also be written as a **while** construct.   ◁

8.1.4 Vectorizing a Loop

In the introduction to loops, you saw that looping is almost always the more
unelegant solution, and can often be avoided. As S-PLUS is vector–oriented,
it is typically possible to rewrite a loop into a shorter and more readable
code. The example below shows how expressions can be rewritten to avoid
a loop completely. We calculate 1000 values at once and check if we have
already fulfilled the criterion to stop, that is, if the sum over these values is
larger than 1000 (which is pretty likely, as we sum up the values 1 to 1000).

```
> n <- 1:1000                       # initialize 1000 values
> s <- cumsum (n)                   # calculate cumulative sum
> # The smallest value for which s is larger than 1000:
> s [s > 1000] [1]
        1035
> # and the corresponding n
> n [s > 1000] [1]
        45
```

For large loops, such an approach saves time and is easier to read. On the other hand, you need to have a good idea how many iterations it might take.

8.2 Writing Functions

Just like a macro in other languages, a function is simply a sequence of commands put together into a single new command, the function's name. In contrast to macros, S-PLUS functions offer the same flexibility and capability as other modern programming languages, like C or Pascal. Functions have arguments passed to them, and results are returned to the calling routine. All other variables defined in the function's body are kept internally and disappear after the function is executed.

In S-PLUS, a function and its usage are very similar to their usage in mathematics. In mathematics you write $y = f(x)$, and the corresponding statement in S-PLUS is

```
> y <- f(x)
```

To demonstrate a function's principle, we write a short function named **cube** which calculates the third power of arguments supplied, and returns this result.

```
> cube <- function (x) { return (x^3) }
```

This function can, after its declaration, be used like any other S-PLUS function. It cannot be distinguished from system functions except by its location (it is stored in a different directory). It is also permanently stored, such that it will still be there if you quit S-PLUS and start it again (see Section 10.1 for more details).

This function now inherits the full flexibility of all other S-PLUS functions just because it is handled by the S-PLUS interpreter. It can not only calculate a single cubed value like in

```
> cube (3)
        27
```

but you can also supply a vector,

```
> x <- 1:5
> cube (x/2)
        0.125 1.000 3.375 8.000 15.625
```

call it recursively,

```
> cube (cube (x))
        1 512 19683 262144 1953125
```

or pass a structure like a matrix.

```
> x <- matrix (1:4, 2, 2)
```

```
> x
        1 3
        2 4
> cube (x)
        1 27
        8 64
```

A function can also have more than one argument (or none at all). We will imitate the system's division *function*. Yes, the operand "/" in S-PLUS is itself a function, which can be accessed and *modified* exactly like any other function. Enter `get ("/")` to see the declaration.

The following function simply divides the first argument supplied by the second.

```
> divide <- function (x,y) { return (x/y) }
```

Here are some examples of calls to the new function `divide`.

```
> divide (7,4)
        1.75                          # two numbers as arguments
> a <- 5:10
> b <- 20:25
> divide (b,a)                        # two vectors as arguments
        4 3.5 3.14286 2.875 2.6666 2.5
> divide (1:10,3)                     # a vector and a number
        0.3333 0.6666 1.0000 1.3333 1.6666 2.0000 2.3333
        2.6666 3.0000 3.3333
```

The more general syntax of a function declaration is as follows.

---

function-name <- function (*arguments*)
{
  function body (S-PLUS *expressions*)
  return (*output arguments*)
}

The expressions in italics have to be replaced by valid names and expressions. The *arguments* are just a listing of parameters to be used inside the function. The *expressions* can be any valid S-PLUS expressions, which are evaluated as S-PLUS executes them, and the *output arguments* can be any data or variable known to the function.

Note that variables you want to use in the function do not need to be declared in advance like in C or Pascal, S-PLUS does it for you when it comes across an assignment.

It is important to know that S-PLUS searches for a referenced variable, and if it does not find it in the function's scope, it searches for it in the interactive environment. You can access a variable defined interactively from inside a function, but for consistency it is better to pass the variable as a parameter to the function.

| Note | Instead of using `return` at the end, you can also simply put a variable's name as the last expression. S-PLUS returns the output of the last expression in the function, but `return` is the more proper way of returning values. Using this practice assures that the return argument is indeed the one wanted. In this fashion, a declaration like

```
> f <- function (x) x^2
```

is valid, but not as proper as using `return (x^2)`.   ◁

| Note | If you do not want to give back anything at all, or do not want it to be printed, use the `invisible` command.

The expression `return (invisible())` at the end of a function causes nothing to be returned at all, nor will anything like NULL be printed on the screen. The expression `return (invisible(x))` returns the contents of the variable x, but does not show it on the screen.   ◁

8.2.1 Scope of Variables

We now cover some of the basic differences between functions, which are used by S-PLUS and other languages, and macros, which are still very common in statistical software.

To begin with, note that all variables declared inside the body of a function are local and vanish after the function is executed. Their scope (the environment where the variables are known) does not extend beyond the function's limits. To demonstrate this effect, we use the following example. Pay attention to the value of z before and after execution of the function f.

```
> z <- 123
> f <- function (x, y) { z <- x/y; return (z) }
> x <- 3
> y <- 4
> z                              # z is not modified by
      123                        # declaring a z inside f
> f (x, y)
      0.75
> z                              # z is also not modified
      123                        # by executing f
```

Making local variables into global ones, such that their names and values appear in the interactive environment, is not advisable because unintended or uncontrollable side effects can occur, such as overwriting another variable. The only proper way of handling local variables is to return them from the function using `return`.

In some extreme cases, you might want to assign values from inside a function to the interactive environment. The function `assign` is useful for this.

```
> assign ("x", 123, where=1)
```

assigns the value 123 to the variable x in the interactive environment. The argument `where=1` tells the system to assign to a variable in the first entry of the search path, which is typically the interactive environment or the data directory, respectively. The second argument is any S-PLUS expression that can be evaluated, such that

```
> y <- 123
> assign ("x", y, where=1)
```

does the same as the statement above.

Note The function `assign` will overwrite already existing variables without confirmation, as its effect is like assigning something to a variable using the `<-` operator.   ◁

The `assign` statement can be abbreviated: The operator `<<-` assigns the righthand side of the expression to a variable defined on the lefthand side. We continue the example from above and write

```
> x <<- 123
```

to assign 123 to x in the global interactive environment from within a function. Understand that this way of assigning variables globally is only an exceptional solution.

8.2.2 Parameters and Defaults

A function can have many arguments. In such cases it is useful if you do not always have to enter all the parameters. In this situation default settings for parameters are useful. In S-PLUS functions, you can assign a default value to a parameter during the declaration using the = sign. Then, if a function is called without explicitly specifying a parameter, its default value will be used.

Example 8.5. A function with default parameter settings
The following function assigns default values to its parameters.

```
f <- function (x=1:10, y=(1:10)^2, showgraph=T)
{
    if (showgraph) { plot (x, y) }
    else { print (cbind (x, y)) }
    return (invisible())
}
```

If a function is declared as in the example above, x has the default value 1:10, and y the default value (1:10)^2. The parameter showgraph is set to T (TRUE) by default. Thus, all the following commands produce the same result.

```
> f (1:10, (1:10)^2, T)
> f (1:10, (1:10)^2)
> f (1:10)
> f (y=(1:10)^2)
> f (showgraph=T)
> f ()
```

Moreover, the parameters can use other parameters in the default settings. See the next example and note that y is a function of x in the declaration of y, and that error is a function of both x and y.

Example 8.6. A function with recursively declared parameters
The parameters of the function declared use other parameters to set default values.

```
f <- function (x=1:10, y=x^2, error=(length(x)!=length(y)))
    {
        if (error) return ("Lengths of x and y do not match.")
        else return (cbind (x, y))
    }
```

⎡Note⎤ S-PLUS uses the so-called *lazy evaluation mechanism*. This means that expressions are evaluated when needed, not when declared. In the example above, if x is changed before y is used for the first time, y no longer takes the square value of the default x, but of the latest assigned value of x.   ◁

⎡Note⎤ If you want to see a function's declaration instead of executing it, enter the function's name without parentheses (), just like you do to see a variable's value.   ◁

Many S-PLUS functions are written in the S-PLUS language. This offers the advantage that you can have a look at the complete source code (see the functions mean or hist) and even modify them!

8.2.3 Passing an Unspecified Number of Parameters to a Function

In a function declaration, you can define the argument, ..., in any position of the parameter list. As a consequence, the function is able to accept any number of arguments (like the S-PLUS functions boxplot and c). In the function body, the ... argument has to be evaluated "manually."

The usage of a function is different, depending on where the ... appears. Here are two simple examples.

```
> f1 <- function (x, ...) { some commands }
> f2 <- function (..., x) { some commands }
```

In the first case, where x appears before the ..., f1 can be called by

```
> f1 (3)
```

and x is assigned a value of 3 within the function. But if you call

```
> f2 (3)
```

the value 3 is a part of ..., and x has no value at all. You need to specify

```
> f2 (x=3)
```

to obtain the same result (that x gets assigned the value 3 in the function). Sometimes, like in the boxplot function, it is useful to have the ... in the beginning, followed by optional arguments for special purposes. On the other hand, it is not useful to put required arguments behind the ... parameter. The following example shows how ... arguments are treated.

Example 8.7. Usage of ... as a function argument

The following function shows how to treat an unspecified number of parameters using the ... expression. We compare the distributions of the data passed to the function. To do this, we calculate the minimum and maximum over *all* data and set the x-axis to the same boundaries for all histograms.

```
f <- function (..., layout=c(3, 3))
{
    L <- list (...)                    # convert data into list
    total.min <- +Inf                  # initialize the minimum
    total.max <- -Inf                  # and the maximum
    for (i in 1:length(L))             # and loop over all data
    { total.min <- min (total.min, min (L[[i]]))
      total.max <- max (total.max, max (L[[i]]))
    }
    par (mfrow=layout)                 # set up graphics layout
    for (i in 1:length(L))             # create the graphs
    { hist (L[[i]], xlim=c (total.min, total.max)) }
    return (total.min, total.max)
}
```

After converting the ... values to a list, you can do anything with this variable. Note that list elements are accessed with double brackets. We will have a detailed look at lists in Section 8.7.

You can think of many other function applications that can handle any number of parameters passed to it.

8.2.4 Testing for Existence of an Argument

To avoid errors, or at least to react accordingly if a required parameter were not passed to a function, you must check if a function parameter is missing. You can use the `missing` function to check if an argument was actually supplied to a function. The returned result is a logical value, TRUE if the parameter asked for is missing, and FALSE if it was supplied in the call. A typical line of code *inside a function* to check whether a parameter x was actually passed to the function or not, is as follows.

```
if (missing (x)) print ("x was not specified.")
```

If a variable x of a function is set per default, and not changed from the call, the output of `missing (x)` is still T (TRUE).

8.2.5 Using Function Arguments in Graphics Labels

If you have ever written a function that creates graphs, you know it is often de-sirable to use the variable's name supplied as an argument in the graph's title. For example, calling `my.funny.graph (my.favorite.data)` should produce a plot of the data with the title "a plot of my.favorite.data."

To do this, you need to store a string containing exactly what was passed to the function, as in the following example.

Example 8.8. Retrieving call arguments inside a function

The following function, f, prints the argument exactly as it was passed. Declaring f as

```
> f <- function (x) { return (deparse (substitute (x))) }
```

produces the following results when calling f with different arguments.

```
> f (2)                          # pass a number
    "2"
> f (x=2)                        # pass a number assigned to x
    "2"
> f (x)                          # pass a variable name
    "x"
> f (sin (pi))                   # pass an expression
    "sin(pi)"
```

In this way, you could add a title to a graph containing what was passed as x.

```
> title (deparse (substitute (x)))
```

If a function has more than just one parameter, the x in the expression deparse (substitute (x)) can of course be replaced by any other parameter name. Another useful tool is the functon sys.call(), which returns the complete call to the function. The returned data is a list: the first element is the name of the function, the second element is the first argument, etc.

8.3 Debugging: Searching for Errors

Now it is time to write your own functions. If you haven't tried this yet, don't worry. You'll have plenty of opportunities in the exercise section. Of course when we write functions, they don't always do what we want. This is the trouble with computers in general. They do exactly what we tell them to do, but this is very often not what we want them to do. This section is about searching for sources of errors.

We will deal with four types of errors.

- Syntax errors.
  While the system is reading the definition of a function, it encounters a problem and stops reading the definition. An example is to close more brackets than we have opened. We are given an error message, typically beginning with Syntax Error: ...
- Invalid arguments.
  The function was passed arguments that do not allow it to correctly execute the function. Calling a routine to do a calculation on numbers and passing character variables to it causes such errors.
- Execution errors.
  While executing the function, S-PLUS encounters an error like an undefined operation. One such example is plotting x against y with x and y having different lengths. S-PLUS stops the execution and we need to track where the error source comes from. Most of the time, S-PLUS tells us
  Error in ...
  If the execution was aborted in a statement that is not the origin of the error, like when we try to plot only missing values, we need to find the source with tools referenced in the following item.
- Logical errors.
  The S-PLUS system completes the operation without detecting an error. We get a result returned which is wrong, like the mean value of a data set being smaller than the smallest value. We need to track the whole function for a logical error that caused a programming error.

The list of problems and errors above is, in general, in ascending order of difficulty of finding the source of the problem. We will now tackle them systematically, introducing the tools S-PLUS has to handle them.

8.3.1 Syntax Errors

Syntax errors are detected most easily. We might have forgotten a closing bracket or confused the name of a function with something else, and the system tells us *while reading in the declaration*, that it disagrees with what it just read.

```
Syntax error: Unbalanced parentheses,
expected "}", before ")" at this point:
```

Such a message is typically followed by the line of code in question, and having another look at the code should clarify what the problem was.

| Note | If S-PLUS gives a syntax error when reading something in from a file, and you cannot see directly what the problem was - maybe because you cannot find the line of code - mark the code using the mouse, and copy and paste it into the command window. S-PLUS will react on the input directly and you can see where the problem is coming from.   ◁

When executing a longer sequence of commands, for example if source (*"filename"*) is used, it helps to set

```
> options (echo=TRUE)
```

To change S-PLUS's sensitivity to warnings, you can change the option warn to the values from -1 to 3. Setting

```
> options (warn=3)
```

lets S-PLUS terminate the current process whenever a warning occurs. By default (which is options (warn=0)), warnings are collected and returned at the end of a statement's execution.

8.3.2 Invalid Arguments

When writing a function, some errors can be checked for automatically. For example, if you wrote a function that gets an input parameter x, and x is not allowed to be negative, you can check for it and exit the function in a controlled manner by issuing an error message.

```
f <- function (x)
{
   if any (x < 0) stop ("x should be greater than 0")
   ...
}
```

The function **any** returns TRUE if any of the logical expressions is true, and FALSE if all of the logical expressions are false. Another function, **all**, returns TRUE if all logical values are TRUE, and FALSE otherwise.

The function **stop** prints the message inside the brackets and lets the function exit immediately. **warning** is a less crude function that issues a warning message, but continues to execute the function. Replace **stop** in the example above with **warning** and the body of the function is executed, but a warning message is printed if x is less than zero.

You can also use **return** anywhere in a function. If S-PLUS encounters a **return** statement, it returns the specified value and exits the function, no matter where you are inside a function's body. We have already discussed function writing and the use of **return** in Section 8.2.

8.3.3 Execution or Runtime Errors

The problem with runtime errors is that they occur while a function is executed. Your function might have called other functions, such that you do not know where the crash in fact occurred. Determining the line of code where the function was actually aborted is the topic of this section. In the good old days, we used to insert **print** statements everywhere in the function code and execute it again, such that we could search for the last **print** statement executed before the crash occured.

If the function was terminated and you were informed that S-PLUS dumped something, like

```
Error: naxy2: Vector of all missing values
Dumped
```

a direct approach would be to call the **traceback** function. **traceback** hierarchically shows all the function calls that were executed when the function crashed.

```
> traceback()
Message:
4: eval (zz, caller)
3: plot.xy ("plot")
2: plot.default (NA, NA)
1:
```

We see that the crash occurred when **plot.default** was called with two missing values as arguments. **plot.default** called **plot.xy**, which in turn called **eval**, and this is where the crash occurred. Our line of code to examine is **plot.default**.

Now we know where the error occurred. If we see the problem directly, fine. If not, we might want to examine the variable contents during the function's execution, or get some more information on the other commands being processed. This is covered in the following section.

8.3.4 Logical Errors

We will now examine advanced S-PLUS debugging, pursuing the contents of variables during a function's execution and more. If you are familiar with other programming languages like C or Pascal, you might know that these languages typically come with a debugger. The purpose of a debugger is to trace the state of a program while it is being executed. To do this, the program is not executed all at once, but interrupted in certain places or executed line by line. S-PLUS has these facilities, too.

If we know where a problem occurs but are unsure what the problem is, we might want to interrupt the program in that specific place and look at the contents of the variables. This is done using the **browser** function. We can insert the statement, **browser()**, anywhere inside a function. Every time S-PLUS encounters this statement while executing a function, it stops and prompts the user for interaction. The commands and the syntax duplicate what we are used to in the interactive S-PLUS session. We can look at the contents of variables after only a portion of the function was executed. It goes even further than that. We can also overwrite any variables defined in the function by assigning something to them. If the browser is quit entering a 0, the function will use the newly assigned value for the variable. We can also declare functions, call a plot, and more. For most applications, this might be the most convenient tool.

Table 8.1 lists the basic browser and debugger commands. We will take a look at the **debugger** now.

Table 8.1. Some **browser** and **debugger** commands

Command	Effect
?	Show the names of all variables currently known (inside a function, for example)[1]
0	Browser: exit, Debugger: go back one level
1, 2, 3,...	Show the contents of variable no. 1, 2, 3, etc., as listed when entering ?
any S-PLUS *expression*	Gets executed and results or assignments are kept within the current scope of variables

[1] *Global* variables defined in the interactive environment are also known in any function.

Building on the previous section, another facility for debugging goes back to the level of function calls and lets us look at the state of the system when the function crashed. After the error occurs, but before we are thrown back to the interactive environment, all function calls and variables - together with their current values - are stored. As this might take some time, depending on the complexity of the function, the system default when an error occurs is to dump only the levels of function calls. By default, the settings are

```
> options (error=dump.calls)
```

If we want to go through the levels of function calls and examine the variables' contents, we need to change the setting to

```
> options (error=dump.frames)
```

and run the function again. The dump will then be done (if the function crashes again), and we can inspect the state of the system by using the debugger.

```
> debugger ()
```

Finally, if you need the functionality of a full blown debugger, the inspect function is the tool of choice. inspect keeps everything under control, such that we pass what we want to execute to inspect, which, in turn, executes this expression and guides us further through the debugging process. We can execute the next expression, enter another function called or skip over it, execute S-PLUS expressions, and more. If we find the point of the function crash, we are still in the inspect environment and can examine everything. As an example, we have a function f and call it with the arguments x and y. As f (x, y) crashes somewhere, we inspect what happens.

```
> inspect (f (x, y))
```

Since inspect offers so much functionality, we might want to restrict ourselves to the basic commands quit and eval. The quit command is used, obviously, to quit. If you need to evaluate an expression, for example for printing out a variable's contents, you can enter an eval *expression*, like in eval x, to print out the contents of x. The step command executes the next step, which is always shown before actually executed. If you need more functionality, entering help at the command prompt of inspect shows a table of the inspector's full functionality, which is easy to understand.

As you can see, the roots of S and S-PLUS are in the computing area. We currently do not know of any other large statistics package offering debugging facilities like these. The remainder of this section presents more evidence of this.

8.4 Output Using the cat Function

In addition to the simple print function, which prints a variable's value, there is a much more powerful tool: the cat function. If you know the cat function from the C language, you will find many similarities. Here are some examples.

```
> x <- 1:3
> cat (x)
      1 2 3
```

```
> cat ("Hello.")                        # the world is ignored here
        Hello.
> x <- 7
> cat ("x has a value of", x, ".")
        x has a value of 7.
```

S-PLUS and C have similar control characters (see Table 8.2). These control characters can be used in any string, and the S-PLUS functions will deal with them accordingly.

For example, a graph title going over two lines can be created by inserting a new line control character somewhere in the title string and entering `title` (`"First\nand second line"`). Note that there is no space around the \n, as this would add an extra space to the title.

Table 8.2. Control sequences in S-PLUS

Control Sequence	Effect
\n	New line
\t	Tabulator
\\	Backslash (\)
\"	"
\´	´
\#	#
\b	Backspace
\r	Carriage return
\octalcode	Octalcode is a number coding the character in octal format (not corresponding to the ASCII code, see a PostScript manual)

Other applications can be formatted using tabulators or special characters in hexadecimal code.

8.5 The paste Function

The `paste` function is one of S-PLUS's most universal functions. It takes any number of arguments and concatenates them into strings. Some impressive examples are given below.

— title strings for graphics

```
> s1 <- "Plot of x and y"
> s2 <- paste ("produced on", date ())
> title (s1, s2)
```

– names for matrix and array rows and columns

Assume that a matrix x with some medical data exists.

```
> row.names <- paste ("Patient", 1:nrow(x))
> col.names <- paste ("Variable", 1:ncol(x))
> dimnames (x) <- list (row.names, col.names)
```

assigns labels to rows and columns of the data matrix like in

	Variable 1	Variable 2
Patient 1	114	72
Patient 2	121	78

The general syntax is

```
> paste (..., sep=" ", collapse=NULL)
```

sep is the field separator, which is set to " " by default. A field separator is put in the space between two concatenated objects.

The argument collapse is the object separator, empty by default, such that the objects to be pasted together do not get collapsed. If collapse is modified to "", the result is a long string instead of a vector of strings, if the arguments were vectors.

8.6 Elements of Object-Oriented Programming

Object-oriented methods are very useful for many applications. In fact, most of the S-PLUS system is programmed in an object-oriented way, often called Object-Oriented Programming Style, or OOPS. If you are going to work on a problem involving different types of data treatment, consider using an object-oriented approach. As we will see later on, one of the advantages of the OOPS way of problem solving is that it generates a very general approach to the problem. It enables the use of modules for objects and functions that are generated, and keeps the system extendable without having to modify any of the previously written lines of code.

There are many, many classes of objects in S-PLUS. You might have used them for quite a while already without taking explicit notice. Using functions and objects in an object-oriented environment does not necessarily require an understanding of the underlying ideas.

Some examples of pre-defined classes in S-PLUS are the classes matrix, numeric, and lm (for linear models). A variety of class functions have been written for these objects. If nothing special is required, there is always the default function, which gets executed when no special class function is defined. On the other hand, if you want to deal with objects of a specific class in a specific way, you can always extend the function library by adding a class-specific function to it.

We can get an idea of how many pre-defined classes there are in S-PLUS by looking at the number of class-specific print functions. Have a look at the definition of the print function by entering

```
> methods ("print")
```

to see a list of class-specific print functions.

Let's look at an easy example. As you have seen before, the system-supplied function mean returns a scalar (a single number) if applied to a vector or a matrix. On the other hand, if mean is applied to a list, an error message is issued saying that data of mode list cannot be input to the mean function. In the following, we will extend the function to return the value appropriate for the class of the object supplied to it:

– For a vector, a single number, the mean of the vector elements
– For a matrix, a vector of values, each being the mean over one column
– For a list, a vector of values, each being the mean over one of the components of the list

To avoid confusion, we will name our function mmean. The result, the mean value or the vector of means, will be assigned a new object class, which we will name mmeandata.

Our idea is to implement a class library in S-PLUS to deal with this problem. For this special purpose we are now going to write a collection of routines dealing with the different types of data that could be passed to the function, like simple vectors, matrices, and lists. As we go along, you will learn about the most elementary functions that deal with object-oriented data, such that another class of objects can easily be implemented.

The general function mmean is a sort of filter. It simply checks to which class the passed object belongs, and passes everything unmodified to the appropriate subclass function to handle. Note that no matter which class the data belong to, the function called will always be the same.

Every object-oriented function library typically has such a high level filter. In S-PLUS, we can use the function Usemethod to check the type of the argument passed, and the appropriate class function is automatically called. We go ahead and define our most global function:

```
> mmean <- function (x, ...)
+ { UseMethod ("mmean") }
```

The ... argument in mmean is put there in order to pass special parameters. This might be necessary if, for example, the data contain missing values. If the function mmean is called together with the argument na.rm=T, mmean passes this argument on to the subsequently called function.

If a data set of class matrix is passed to mmean, the function Usemethod searches for a function mmean.matrix, in order to correctly treat the matrix. If it does not find the function mmean.matrix, it searches for a function of the name mmean.default and executes this function. If both cannot be found,

an error message is issued. What is left to do is define the default function
plus some class-specific functions.

```
> mmean.default <- function (x, ...)
+ {
+   result <- mean (x, ...)          # call the standard
+   class (result) <- "mmeandata"    # mean function
+   return (result)                  # and assign a class
+ }
```

Note that in the end, the returned data, `result`, is assigned the class
`mmeandata` using the `class` function. In the same way, data can get any
class assigned using the same function. A class assignment does nothing but
attach a string object to an S-PLUS object. Here is a short example. We
assign and remove a class attribute to a variable x, and after each step we
check what class x belongs to.

```
> x <- 1:5
> class (x)                  # check of what class x is
        NULL
> class (x) <- "any.class"   # assign a new class
> class (x)                  # check the class again
        "any.class"
> unclass (x)                # remove the class from x
> class (x)
        NULL                 # and check again
```

Now we write a specific `mmean` function for the class `matrix`, which will be
stored as `mmean.matrix`. This function calculates the mean of each column
of a matrix.

```
> mmean.matrix <- function (x, ...)
+ {
+   result <- apply (x, 2, mmean.default, ...)
+   class (result) <- "mmeandata"
+   return (result)
+ }
```

As before for the class `matrix`, we write a specific `mmean` function for the class
`list`, which calculates the mean for each element of the list. As the function
`lapply` does exactly this, we `unlist` the result from `lapply`, in order to
obtain a vector instead of a list of values.

```
> mmean.list <- function (x, ...)
+ {
+   result <- unlist (lapply (x, mmean.default, ...))
+   class (result) <- "mmeandata"
+   return (result)
+ }
```

Now we need to extend the existing libraries of **print** and **plot** functions in order to treat our new object class, mmeandata, the way we want. We do not need to define a default **print** function - S-PLUS has done this for us. Look at the definition of the function **print** to check this out.

Here is the new class-specific **print** function for our object class mmeandata. It does nothing special, just reveals that the data was generated by the mmean function.

```
> print.mmeandata <- function (x, ...)
+ {
+   cat ("This data belongs to the mmean class.\n")
+   print.default (x)
+ }
```

Now note that we use the default print method to print out the values. Can you think of what you would do to have different **print** functions for each class that could be supplied to the mmean (class of) functions?

Note Recall again what we just did. We extended the system's **print** function by adding to it a new behavior for a new class, but we did not even touch the **print** function itself. We just added new functionality to it by defining the behavior for a new class.

If you are working in a multi-user environment, this is a way of developing a completely independent general function set. New enhancements (adding new classes of objects) do not need to touch the old functions, which saves us from introducing new errors.   ◁

Finally, we can also enhance the behavior of the **plot** function to do something special with our new object class. We want to plot the mean values in different ways and create a 2x2 layout with different graphs in each picture, in order to compare the mean values of each element of our data, whether it is a matrix or a list.

```
> plot.mmeandata <- function (x, ...)
+ {
+   par.old <- par()
+   on.exit (par (par.old))
+   par (mfrow=c(2, 2))
+   plot.default (x, type="h")      # use default function
+   # plot all data points at the same x coordinate x=1
+   plot.default (rep (1, length(x)), x)
+   boxplot (x)                     # show a boxplot
+   hist (x)                        # and a histogram
+ }
```

We did not try to generate objects of the classes for which we just defined methods. Go ahead and try them out. Create an object, assign a class to it, and run the functions we just defined.

Note that classes are inherited by operations. If a vector x belongs to a certain class and you define y <- x^2, then y automatically inherits the class to which x belongs.

You can find more on all of these mechanisms in the manuals, but this should be enough to start off. Now you have a better understanding of what is going on inside S-PLUS when you call certain functions like print.

You should go ahead now and create a vector and a matrix, assign classes to them, and try out our newly created functions (if you haven't done so already).

8.7 Lists

A list is typically a very flexible structure in which to store data. Its main usage is in situations where some data belong together by their contents, but the structure is very heterogenous. Imagine a data base of somebody's floppy disks, each disk with a number, a short text description of the contents, and a list of files it contains. It would not be a good idea to store this kind of data in a matrix or data frame. A list would be a good structure to store all these data in one variable.

The main characteristic of a list is that it has, at the top level, pointers to objects that can be of any type. For S-PLUS, this results in a structure that can store any type of object in a single variable. In contrast to matrices, vectors, or data frames, the objects in a list are completely independent, such that they can be of different types and sizes. For example, a list can consist of a vector, a matrix, and another list.

A list is constructed by calling the function list and enumeration of the objects to go into it. Here is an example for creating a list with three elements, each of them being a vector.

```
> L <- list (1:10, c(T, F), c ("Hey", "You"))
```

This list has three elements: the first element is a vector containing the numbers 1 to 10, the second element contains the Boolean values TRUE and FALSE, and the third element contains two strings. Of course you can create a list in many different ways. Here is how we could have created the same list using variables:

```
> x <- 1:10
> y <- c (T,F)
> z <- c ("Hey", "You")
> L <- list (x, y, z)
```

The list, L, is printed in the same manner as any other structure.

```
> L
    [[1]]:
        1 2 3 4 5 6 7 8 9 10
    [[2]]:
        T F
    [[3]]:
        "Hey"   "You"
```

Looking at the output display, you can directly see how list elements are addressed.

```
> L[[2]]
        T F
> L[[3]][2]
        "You"
```

┌─────┐
│ Note │ Keep in mind that list elements are addressed by double square brack-
└─────┘
ets [[ ]]. Single brackets refer to array-like structures.   ◁

Elements of lists are assigned, accessed, and overwritten in the same way. The next section deals with extending and shrinking lists.

8.7.1 Adding and Deleting List Elements

To add a new element to a list, simply put it into a location that is not already used. The list L, from above, consists of three elements that can be determined in the usual way by

```
> length(L)
        3
```

You can add a new element by assigning the new element to an unused index.

```
> L[[4]] <- c ("my", "new", "element")
```

You can use any index you like, but using the next free index is preferable, as the elements in between are assigned the value, NULL. If you add element number 99 to a list that has 3 elements, you automatically create elements numbered 4 to 98.

A flexible command to use would be

```
> L[[length(L)+1]] <- next element
```

where *next element* can be any type of data. Note that the new element of L is added directly after the last element, without explicitly knowing the length of the list.

If an already existing index is specified, of course the element is overwritten. You delete an element by setting its value to NULL. The command

```
> L[[2]] <- NULL
```

deletes the second entry of the list, such that it does not appear anymore, not even with its newly assigned "value," NULL. It is currently impossible to remove several elements of a list at once. Statements like L[[1:2]] <- NULL are not executed. Similarly, it is impossible to use negative indices (like in vectors or matrices) to express suppression of an element.

| Note | After deleting an element, all remaining elements of the list are shifted - their indices are decreased by 1!   ◁

The idea from the above note can be used to write an interesting for loop. Think about what happens on the following two input lines and then try it out.

```
> L <- list (1:10, 11:22, 44:24, "string", T)
> for (i in 1:4) { L[[2]] <- NULL }
```

The re-enumeration of list elements after deleting an element is sometimes a source of problems. The following little program is intended to delete the second, third, and fifth element of the list L. Do you see what is wrong with it and what the program really does? Try to guess the result before executing it in S-PLUS.

```
> L <- list (1:11, "a", "b", "c", "d", "e", list (T, 2:5))
> L[[2]] <- NULL
> L[[3]] <- NULL
> L[[5]] <- NULL
```

| Note | The *safest* way to delete elements from a list is to start at the end. Delete the highest index first, then the second highest, etc., and the smallest last.   ◁

8.7.2 Naming List Elements

Similar to objects like vectors or matrices, lists can receive names to describe their elements. The function to use is names (not dimnames, like for matrices). The list elements can be accessed by their names, which makes the structure more self-explanatory. As we will see, the names can be changed at any time. Here we create a list as before, and assign names to the elements while defining it.

```
> L <- list(number=1:10,bool=c(T,F),message=c("Hey","You"))
> L
    $number:
        1 2 3 4 5 6 7 8 9 10
    $bool:
        T F
```

```
$message:
      "Hey"   "You"
```

The same result can be obtained if the names are assigned to the elements after having created the list.

```
> L <- list (1:10, c (T,F), c ("Hey","You"))
> names (L) <- c ("number", "bool", "message")
```

Now the elements of the list can be addressed using their names. This is especially useful if you do not want to have to remember the indices. The handling is exactly like with indices because the names are matched to them. We retrieve the element bool from the list L by entering

```
> L$bool
      T F
```

You make an assignment to an element whose name is not already an element name, another element with the new name is added to the list. The command

```
> L$comment <- "we assign a new element"
```

adds a new element of name, comment, to the list.
For modifying existing names, the old names can simply be overwritten using the names function.

```
> names (L)                         # the names of the elements
      "number"  "bool"  "message"  "comment"
> names (L)[2] <- "true and false" # new name for element no. 2
> names (L)                         # the new names
      "number"  "true and false"  "message"  "comment"
```

8.7.3 Applying the Same Function to List Elements

The tool lapply is designed for working on all elements of a list using the same function. For example, you might want to calculate the mean of every list element. We first create a numeric list, and then use lapply to show how it works.

```
> L <- list (vec=1:10, mat=matrix (99:88, 3, 4))
> lapply (L, mean)
    [[1]]:
        5.5
    [[2]]:
        93.5
```

Remember that all standard arithmetic is also a function within S-PLUS. So we can also subtract 1 from each element of the list.

```
> lapply (L, "-", 1)
   [[1]]:
       0 1 2 3 4 5 6 7 8 9
   [[2]]:
           98  95  92  89
           97  94  91  88
           96  93  90  87
```

The small example shows that `lapply` takes any number of arguments and passes them unmodified to the function that is to be applied to the list.

| Note | If a list becomes large, when adding or deleting an element of the list, the list is copied into memory, modified and then stored (if no error occurred). This can consume a lot of memory, but guarantees that the previous state of the system can be recovered if an error occurs during execution of the command. If you create an empty list first, this memory problem is reduced, as no further copy of the list in memory is necessary if an element is added or deleted. An "empty" list with ten elements can be created using

```
> L <- vector ("list", 10)
```

This creates a vector of mode list with 10 empty elements. Later on, you can assign values to the different elements of L.   ◁

8.7.4 Unlisting a List

Sometimes it is necessary to extract the elements of a list to compose them again into a single structure. To do this, the function `unlist` takes all elements of the list and combines them in the most global data structure. For example, a list with vectors and matrices with numeric contents becomes one long vector containing all the numbers; a list with numeric and character contents becomes a character mode variable.

If we continue the example from above, applying the `mean` function to all elements of a list, we might want to have a vector of means instead of a list. The function `unlist` does exactly what we want.

```
> unlist (apply (L, mean))
     5.5 93.5
```

The following exercises will show in more detail what we just discussed.

8.8 Exercises

Exercise 8.1

Write a function, `lissajous`, that accepts the parameters a and b and plots the corresponding Lissajous figures. Set meaningful default values for a and b. For a definition of Lissajous figures, refer to Exercise 5.3, page 77.

Exercise 8.2

Use the functions `outer` and `persp` to plot the following mathematical functions:

$$f(x, y) = 1 - \exp\left(-1/(x^2 + y^2)\right); \ x, y \in [-5, 5]$$
$$f(x, y) = \sin(x)\cos(y); \quad x, y \in [-\pi/2, \pi/2] \text{ and } x, y \in [-\pi, 2\pi]$$
$$f(x, y) = x\sin(y); \quad x, y \in [0, 16]$$

Figure out how to use the functions by looking at the manual or the online help pages.

Exercise 8.3

Create a new class of objects. The class name shall be `data`. The `plot` function for this class shall produce a histogram and a boxplot of the data on a single graphics page, and the `print` function shall return a nice looking table containing the number of observations, the number of missing values in the data, the minimum, the maximum, and the mean value.
Create an object of the class `data` and apply the functions `plot` and `print` to it.

Exercise 8.4

We want to compare data sets from different distributions. In order to do this, write a function `compare.distributions`. The function shall expect a list of data passed to it, and plot histograms of all samples using the same scale and intervals.

Can you let the function determine the layout of the graphs, i.e. the number of rows and columns of the graph matrix?

Try out the function by creating samples from the standard Normal, the t(11) and the t(33)-distribution, with 100 and 10,000 random values, respectively.

Exercise 8.5

Create a list with names, telephone numbers, and addresses of five of your friends. Write a function, `tel`, that gets passed a name and returns the appropriate address and telephone number. This function should handle the case of being passed a name that is not in the list.

8.9 Solutions

Solution to Exercise 8.1

In order to write a function that creates the Lissajous figures, we take the code from Exercise 5.3 (page 81) and write a function around it. The function gets the parameters a and b passed to it, and the third parameter is a default graphics device. Depending on the configuration, the default might need to be changed. Note that we pass the *function name* of the device driver only, and not also the parentheses normally needed to run it (think about it!). What gets passed is a declaration of the function, which is what you see when you enter the function's name at the prompt. The actual execution is done inside the function lissajous.

```
lissajous <- function (a, b, device=graphsheet)
{
    # if no graphics device exists, we open it now
    if (!exists (".Device")) { device() }
    x <- seq (0, 2*pi, length=1000)
    y <- sin (a*x)
    z <- sin (b*x)
    plot (y, z, type = "l", axes=F, xlab="", ylab="")
    title (paste ("Lissajous Curve (",a,",",b,")", sep=""))
}
```

We can go ahead and produce a Lissajous figure by entering, for example,

```
> lissajous (6, 8)
```

Solution to Exercise 8.2

To draw two-dimensional mathematical functions, we use the functions outer and persp. The function outer combines two arrays x and y into a matrix, in which the element [i,j] of the matrix contains the function value f(x[i], y[j]), where f is any function with two arguments.

The function persp draws the surface defined by the matrix. By changing the eye parameter, the viewpoint can be modified. To create a graph, we need to define two vectors (or just one and use it for both axes). The next step is to define the combining function f and to apply it. The result is a matrix whose first dimension equals the length of x and whose second dimension equals the length of y. The matrix itself is stored in z.

```
> x <- seq (-5, 5, length=50)
> f <- function (x, y) { 1 - exp (-1/(x^2+y^2)) }
> z <- outer (x, x, f)
> persp (z)
```

We can do all figures in this fashion, but a more elegant way would be to write a general function that receives the function to draw, the boundaries of the axes, and, optionally, the number of points on both axes.

```
function.draw <- function (f, low=-1, hi=1, n=30)
  { x <- seq ( low, hi, length=n )
    z <- outer (x, x, f)
    persp(z)
  }
```

We can now produce Figure 8.1 by entering the following commands:

```
par (mfrow=c(2, 2))              # to set the graph layout
f1 <- function(x, y) { 1-exp(-1/(x^2+y^2)) }
function.draw (f1, -5, 5)
f2 <- function(x, y) { sin(x)*cos(y) }
function.draw (f2, -pi/2, pi/2)
function.draw (f2, -pi, 2*pi)  # new boundaries
f3 <- function(x, y) { 0.1*x*sin(2*y) }
function.draw (f3, 0, 16, n=40)
```

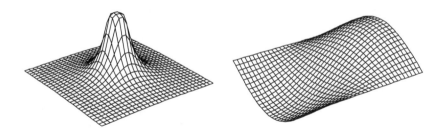

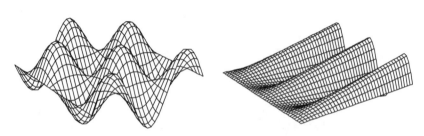

Figure 8.1. Two-dimensional functions drawn with persp.

Solution to Exercise 8.3

Creating a new class of objects requires nothing more than assigning the new class to an object. For example, we create a vector with some data by filling it with random numbers from two different distributions.

```
> x <- c (rnorm (1000, 9, 3), rnorm (1000, 25, 5))
```

We assign the class by using

```
> class (x) <- "data"
```

which creates an object of the new class "data."
Assigning the new class to this object does not affect how S-PLUS will handle the object x, as no functions designed for the new class exist yet. The generic function plot can be extended to handle objects of class data. We will define a new function plot.data that will contain our customized plot commands. The advantage is that now when the normal plot function is called with data of type data, the plot.data function is automatically used. Below is one possible configuration for plot.data.

```
plot.data <- function (x, ...)
{
    par (mfrow=c(1, 2))          # create a 1x2 layout
    hist (x, ...)                # call hist
    boxplot (x, ...)             # call boxplot
    return (invisible())         # return nothing
}
```

Note that we defined a parameter, ..., which can be supplied to our function. We do nothing else with it but pass it on unmodified to the hist and boxplot functions. An application could be to pass a title using plot (x, main="a data object plot"), or you might want to omit the surrounding axes by passing plot (x, axes=F) to the plot.data function.
We can define a print function to handle objects of class data in the same way.

```
print.data <- function (x)
{
    out <- c(length (x), sum (is.na (x)), min (x, na.rm=T),
        mean (x, na.rm=T), max (x, na.rm=T))
    names (out) <- c("observations", "missings", "minimum",
        "mean", "maximum")
    return (out)
}
```

Note that we added the parameter na.rm=T to the calls of min, mean, and max, such that we get these values even if there are missing values in the data (otherwise, the result would be NA). The last thing to do is to call the new functions to see if they work as expected.

Solution to Exercise 8.4

We write a general function that gets passed any number of data sets. The data sets are passed by using the ... argument, and the minimum and maximum over all data sets is determined. The next step is to create a sequence of breakpoints between the minimum and maximum, which will be used as limits of histogram classes, such that every data set is displayed in a histogram with exactly the same layout.

Of course we want to see all the histograms in one window or on one sheet of paper, respectively. We define a parameter to the function, `layout`, which lets the user determine the number of rows and columns for the histogram matrix. If the user does not supply it, we calculate it automatically, such that all pictures fit on one page.

To actually try out our new function, we generate data from a N(0,1), a t(11), and a t(33) distribution, each with 100 and 10,000 sample points, respectively. This gives us 6 samples, which we pass to `compare.distributions`, our newly written function.

```
compare.distributions <- function (..., layout)
{
    data <- list (...)              # convert data into list
    par.old <- par()                # store current settings
    on.exit (par (par.old))         # and restore on exit

    xrange <- range (unlist (data)) # determine min and max
    breaks <- seq (xrange[1], xrange[2], length=20)

    if (!missing (layout))
        par (mfcol=layout)
    else
    {                               # calculate layout
        n1 <- floor (sqrt (length (data)))
        n2 <- ceiling (length (data)/n1)
        par (mfrow=c(n1, n2))
    }
                                    # do the histograms
    lapply (data, hist, breaks=breaks, xlab="")
    return (invisible (breaks))     # return breaks invisibly
}
```

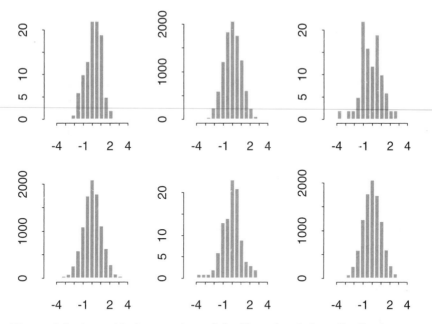

Figure 8.2. A graphical comparison of the Normal and the t-distribution.

Note how the automatic figure layout looks when the value `layout` is not supplied. This setup is for landscape mode. For portrait mode, the two values of `n1` and `n2` need to be swapped.

Finally, we generate the data and pass it to the function, such that we get the result shown in Figure 8.2.

```
> N.100 <- rnorm (100, 0, 1)
> N.10000 <- rnorm (10000, 0, 1)
> t.11.100 <- rt (100, 11)
> t.11.10000 <- rt (10000, 11)
> t.33.100 <- rt (100, 33)
> t.33.10000 <- rt (10000, 33)
> compare.distributions (N.100, N.10000, t.11.100,
+   t.11.10000, t.33.100, t.33.10000)
```

Solution to Exercise 8.5

We define a function, `telephone`, to search in a database for the telephone number and home town of our friends. The database is a list that we will define first.

Look at how the comparison is done. The result of the comparison is stored in a variable which is used later to extract the data from the database.

If you would like to try out more, try to handle the case where there are two friends of the same name found in the database, and extend the function such that a part of the name to search for is sufficient to find the corresponding entry.

```
> # Define the database outside the function
> tel.data <- list()
> tel.data [[1]] <- c("Paul","Tom","Bert","Peter","Carl")
> tel.data [[2]] <- c("007","008","009","010","011")
> tel.data [[3]] <- c("Boston","Seattle","Berlin",
+ "Chicago","Zurich")
> # set names for the list elements
> names (tel.data) <- c("name", "tel", "address")

> # Define the function telephone for doing a name search

telephone <- function (name)
{
    compare <- name==tel.data$name
    # store the comparison for later use as index selector
    found <- (sum(compare) > 0)
    if (!found)
    {
        cat ("Name not found.\n")
        return (invisible())
    }
    # format the output using control characters
    cat ("Name:\t", tel.data$name[compare], "\n")
    cat ("Telephone:\t", tel.data$telephone[compare], "\n")
    cat ("Address:\t", tel.data$address[compare], "\n")
    return (invisible())
}
```

Here are some applications of the function:

```
> telephone ("Tom")
      Name:        Tom
      Telephone:   008
      Address:     Seattle
> telephone ("Fred")
   Name not found.
```

9. Input and Output

This is one of the most important chapters, as it is every user's intention to analyze his or her own data. To do this, the data has to be read into the system before it can be analyzed. This chapter discusses in detail the different ways of reading and writing data.

If your version of S-PLUS contains an import/export facility, this facility is the easiest way to import and export data sets. Importing data in this fashion was described in Section 2.2.1. Writing output from S-PLUS and transferring data files are other important input/output functions covered in this chapter.

As a quick overview, you should be aware of the following commands. To read S-PLUS commands from an external file, use the `source` function. To read data from an ASCII file, use the `scan` function or its alternate, the `read.table` function. To transfer data to and from different S-PLUS systems, use the `dump` and `restore` functions. To write data to an ASCII file, use the `write` function. Each of these topics and functions is discussed in this chapter.

9.1 Reading Data from the Terminal

Entering data from the terminal (keyboard) is easy. Use the simple `scan` function to enter data directly. But remember, there is no way to correct typing errors directly, so this is only useful for short data sets.

S-PLUS prompts you with the index of the next element. You can enter element by element and hit `<Return>` after each entry, but you can also separate data by spaces and occasionally hit `<Return>`. For example,

```
> x <- scan ()
    1 2 3
```

Entering an empty line finishes the input.

9.2 Using the scan Function

The `scan` function is the most universal way to read in all sorts of data. It reads data from the terminal but it is mostly used for reading from a file.

The most important parameters are

```
scan (file=" ", what=numeric(), n, sep, ...)
```

where `file` is the name of the input file, per default the standard input (usually the keyboard), as we have seen in the previous section, and `what` is the type of data to be read in. The default is numerical data. For characters, you can specify

```
> scan ("filename", what=character())
```

or supply a variable of the type desired:

```
> scan ("filename", what="")
```

`n` is the number of observations to read in - as many as there are per default. `sep` can be specified, if the data is not separated by spaces, but by tabulators or other characters.

Here is a simple example. Suppose a file with name data.dat contains the following:

```
1   2   3
4   5   6
7   8   9
```

The command

```
> x <- scan ("data.dat")
```

reads in a vector with 9 elements. We can store it as a matrix by specifying

```
> data <- matrix (x, ncol=3, byrow=T)
```


| Note | If your data file is not in the default directory, you must give the full pathname of its location. Be aware that with Windows, you have to use double backslashes instead of single to specify pathnames. For example, a file in the default directory might be referenced with

```
> scan ("c:\\spluswin\\home\\data.dat")
```

The forward slash used to specify pathnames in UNIX does not pose a problem in S-PLUS, and can also be used to specify a Windows path, as in

```
> scan ("c:/spluswin/home/data.dat")
```

◁

9.3 A Comfortable Function for Reading Data: read.table

The `read.table` function can make your life much easier when trying to read data into S-PLUS. There are a few rules to keep in mind, though, or your table will not appear as you want. The basic syntax of the `read.table` function is simply,

```
> read.table ("filename")
```

This simple specification assumes that the first column contains the row names. If you want the first column to be a part of your matrix, you need to add the option, `row.names=NULL`. The other major point to keep in mind is how the function will handle the first row of the input file for the creation of column (variable) names. By default, the first row is *not* used for the variable names unless it happens to contain one fewer element than the second row of input. In this case, it assumes that the row names are contained in the first column, with no corresponding variable name, and the first row does indeed contain the variable names. This default can be a little deceiving, but can be controlled decisively using the `header=T` (or `F`) option.

The `read.table` function is easy to use when the input file contains only the data of interest, with no variable names, which perhaps contains a mixture of numeric and character data. In such cases, `read.table` automatically determines the type of data in each column and eliminates the `what` option of the `scan` function.

| Note | Labels are typical sources of errors. If a simple label contains spaces, you should put it into quotes or replace the spaces by other delimiters like the underscore or period.   ◁

9.4 Editing Data

At some point you will inevitably make a mistake while typing in your data, which you will want to correct. You can always make a direct substitution of the cell with the incorrect data, as in

```
> x[2, 5] <- 2
```

Sometimes, however, it helps to look at the data to find out where the problem is. In these cases, you need a data editor. There are two basic ways of editing data: using the Object Browser or using a text editor.

Using the Object Browser to edit data was described in Section 2.1.1. Simply select your data and right click to open a dialog box. Selecting [OPEN VIEW] will open a spreadsheet with the selected data which can then be

edited. Versions of S-PLUS without an Object Browser should include the
`data.ed` function, which also opens a spreadsheet editor.

Text-based data editing is done with the `edit` function. The default editor
used with this function (the notepad editor under Windows and vi under
UNIX) may be changed to any other editor using the `editor` option. S-PLUS
has also functions `vi` and `emacs` to edit objects. To change the default editor
for all sessions, it is better to specify the editor using the following

```
> options (editor="my_editor")
```

where the file `my_editor` contains the executable file of your preferred editor
(notepad, vi, emacs, etc.). Once you have settled the issue of which editor to
use, actually editing the data is easy. Simply run the `edit` function on your
data.

```
> x <- edit (x)
```

Notice that we assigned the edited version of x back into the variable x. If
we don't reassign it somewhere, the edited version is lost.

9.5 Transferring Data: The `dump` and `restore` Functions

You can use the `dump` function to write S-PLUS variables to an external file
for future use in S-PLUS. To write data such that they can be used in a
different software package, you should put the data into an ASCII file with
the `write` function (see Section 9.7). The `restore` function then reads the
output of the `dump` function back into S-PLUS.

The function `dump` can write S-PLUS variables of any type to a file. It uses
S-PLUS syntax, such that `restore` or `source` can retrieve S-PLUS commands
directly from the file without any other commands.

Example 9.1. Transferring data with the `dump` function
In this example, we create a variable x and dump it into a file named
`xdump.dat`.

```
> x <- 1:5                  # Create x
> dump ("x", "xdump.dat")   # Dump x into xdump.dat
```

Using any editor to examine the contents of the file `xdump.dat` shows that it
contains

```
"x" <- c(1, 2, 3, 4, 5)
```

This is especially useful for transferring S-PLUS data from one hardware
platform to another. If you want to dump all variables at once, use

```
> dump (objects (), "dump.all")
```

9.6 Reading S-PLUS Commands from File: The `source` Function

The command `source` is intended to read a series of S-PLUS commands from a file, instead of getting the input commands directly from the terminal. If you try out a series of commands over and over again, this is a useful tool.

If you intend to run a batch job (see Section 10.5), use `source` for a short test run.

Under UNIX, the same result may be obtained with the command line `Splus < ` *textfile*, where *textfile* is the name of the file containing the S-PLUS commands. The commands are executed subsequently and the results are treated exactly as if you had entered them from the command line using

```
> source ("S-Plus_command_file")
```

9.7 Writing Text Files

The `write` function is used to write data to an ASCII file such that it can be read by a different software package, or even by S-PLUS using the `scan` function.

The basic `write` function works by simply specifying the S-PLUS object to be written and its intended location. Using the x vector defined in the previous section, a simple example is

```
> write (x, "xdata")
```

which creates a file of name `xdata` with the contents of x. You can optionally specify the number of columns for the output, which may be essential to use for compatability with other packages. The default number of columns used by the `write` function is five for numeric variables and one for character variables. You can use the functions `print`, `format`, `cat`, and `paste` to format output to be written to the files.

| Note | The `write` function writes a matrix into the output file one column at a time. This means that it actually writes the transpose of the matrix. In order to have the output match the original matrix, use the `t` (transpose) function first and also specify the number of columns to use.

```
> write (t(x), "xdata", ncol=ncol (x))
```

You might want to try this out on your own. You should try writing the file, as above, and then reading it back in.   ◁

9.8 Writing S-PLUS Output

To redirect S-PLUS output "as is," the `sink` function is useful. It is a simple way of saving results from a calculation or a specific display into a file to be edited or used later on. The command

> `sink ("filename")`

redirects all the following output to the file *filename*, until the command

> `sink ()`

is used to close the file and direct the output back to the screen.

9.9 Exercises

Exercise 9.1

In this exercise, we will practice reading and writing one of the S-Plus data sets. The object `state.name` contains the names of the fifty states in the United States, and `state.x77` contains various attributes about the states.

a) Use the `write` function to write the populations, incomes, and illiteracy rates to a file named `mystates.dat`.

b) Read the data back into S-Plus, from the `mystates.dat` file, using the `scan` function.

c) Connect the state names to the variables from part a) and write the first ten into a new file.

d) Repeat part b), but now remember that one of the variables is character.

e) Now use the `read.table` function to read the data into S-Plus. Use the function both with and without the `row.names=NULL` option. How do the results compare to each other and how do they compare to the result from part d)?

Exercise 9.2

If you are using S-Plus under Windows and also have Excel, repeat parts a–d of the above exercise using only the import and export facilities in the [File] menu.

Exercise 9.3

We want to track an investment with a fixed interest rate over some years. Develop a function, which takes the parameters *initial capital, interest rate,* and *time in years,* and create a graph that shows the amount of capital over the years. The capital stock changes every year, as we leave the interest obtained on the account.

The trick is that the function should ask the user to supply any of the parameters that have not been supplied to it. The parameters are the initial capital investment, the interest rate, the length of time involved, and the first year of the investment. Be sure to cover the possibility that the user has entered the interest rate as a percentage rather than as a fraction.

This is going to require some extra work on your part, as we haven't covered some of these functions in the chapter.

Hint 1: Use the help facility to find out about the `readline`, `parse`, `eval`, and `missing` functions.

Hint 2: The formula to calculate the capital after n years, with an initial capital x and an interest rate r, is given by $x * (1 + r)^n$. For example, after we invest a capital of \$10 at an interest rate of 10% (=0.1). After one year, the capital becomes \$10 plus \$1 in interest. After two years, we obtain \$11 plus \$1.10 interest, such that the new capital stock is \$12.10.

Using the formula from above, we get $\$10*(1+0.1)^2 = \$10*1.21 = \$12.10$.

In addition, let the S-PLUS function ask for the parameters, if one or more parameters were not supplied by the user.

9.10 Solutions

Solution to Exercise 9.1

It is most convenient to write the transpose of the data (using the **t** function) in question when using the **write** function. We are only going to write the first three columns, which correspond to the variables of population, income, and illiteracy rate.

```
> write (t(state.x77[ , 1:3]), "mystates.dat", ncol=3)
```

Part b) is performed by using the **scan** and **matrix** functions together with the **byrow** option, which will preserve the structure.

```
> partb <- matrix (scan ("mystates.dat"), byrow=T, ncol=3)
> partb
       [,1]  [,2]  [,3]
[1,]   3615  3624   2.1
[2,]    365  6315   1.5
[3,]   2212  4530   1.8
[4,]   2110  3378   1.9
[5,]  21198  5114   1.1
        :     :     :
```

You might want to compare the output you obtain directly with the vector, **state.x77**, just to make sure the processes of writing and reading have worked correctly.

Connect the state names to the state information using the **cbind** function and write the first ten lines to a file (as in part a)).

```
> outc <- cbind (state.name, state.x77[ , 1:3])
> write (t(outc[1:10, ]), "mystate1.dat", ncol=4)
```

When we read in the data this time, we have to keep in mind that they contain a character field and that we have to use the **scan** function's **what** option to accommodate this.

```
> partd <- matrix (scan ("mystate1.dat", what=" "),
+ byrow=T, ncol=4)
> partd
             [,1]          [,2]      [,3]      [,4]
[1,]    "Alabama"        "3615"    "3624"    "2.1"
[2,]     "Alaska"         "365"    "6315"    "1.5"
[3,]    "Arizona"        "2212"    "4530"    "1.8"
[4,]   "Arkansas"        "2110"    "3378"    "1.9"
[5,] "California"       "21198"    "5114"    "1.1"
          :              :          :          :
```

The only problem with the previous solution to reading in the data is that we are left with numeric variables that are stored as character. The `read.table` function is a little smarter and automatically recognizes which variables are numeric and which ones are character. What we have to be aware of with the `read.table` function is how it treats the row names. Recall that if you specify new options, `read.table` assumes that the first column of your data contains the row names.

```
> tablea <- read.table ("mystate1.dat")
> tablea
                 V2     V3    V4
    Alabama     3615   3624  2.1
     Alaska      365   6315  1.5
    Arizona     2212   4530  1.8
   Arkansas     2110   3378  1.9
 California    21198   5114  1.1
```

On the other hand, if you specify the option, `row.names=NULL`, then the first column is assumed to be part of the data.

```
> tableb <- read.table ("mystate1.dat", row.names=NULL)
> tableb
           V1        V2     V3    V4
  1     Alabama    3615   3624  2.1
  2      Alaska     365   6315  1.5
  3     Arizona    2212   4530  1.8
  4    Arkansas    2110   3378  1.9
  5  California   21198   5114  1.1
```

We can easily see the difference between the two methods by simply printing a cell from each of the two resulting matrices.

```
> tablea[2,2]
    6315
> tableb[2,2]
    365
```

In the first matrix, the state names are not considered to be part of the data. In this case the element in the second row, second column is 6315. In the second matrix, however, the state names are part of the data matrix so that the element in the second row, second column is 365.

Solution to Exercise 9.2

Since the creation of data is left up to you and most of the work is done by clicking on different menu options, it is not possible to have a real solution here. Good luck.

Solution to Exercise 9.3

To show the development of our investment over time, we use the formula given and calculate the capital stock for each year. After that, we plot the result obtained. If a parameter is not supplied, the function asks the user to enter it interactively. Figure 9.1 was created by calling interest.rate (100, 5, 15). As a default, we start the investment in 1997.

```
interest.rate <- function (dollars, interest, years,
    first.year=1997)
{
```

Calculate capital development over years. We define an initial investment of dollars, an interest rate of interest, and let it accumulate for years number of years.

The key to the whole solution is the local function, ask, which prompts the user to specify a value, reads the value in as string, and returns it back as numeric.

```
ask <- function (text="?")
{ cat (text)
return (eval (parse(text=readline ())))
}

cat ("\nInterest Rate Calculation.\n")

if (missing(dollars)) dollars <- ask ("Capital : ")
if (missing(interest)) interest <- ask ("Interest : ")
if (missing(years)) years <- ask ("Run Length: ")
```

The interest rate can be given in percent, for example, 5, or in terms like 0.05. If it is greater than 1, we divide by 100 and inform the user.

```
if (interest > 1)
{
cat ("I assume you meant", interest,
    "percent.\n")
interest <- interest / 100
}

capital <- rep (dollars, years+1)
capital <- capital*(1+interest)^(0:years)
```

Now we are ready to create some plots. We create a time series from the data and use the time series plot function (tsplot).

```
capital <- ts (capital, start=first.year, freq=1)
tsplot (capital, col=6 )
title.str <- paste ("Capital development from",
    first.year, "to", first.year+years, "\n",
    "at an interest rate of", interest, "percent")
```

```
        title (title.str)
```

Let us put the result into a string.

```
        result <- paste ("An investment of $", dollars,
            "at an interest rate of", interest,
            "per year sums up to", round (dollars),
            "after", years, "years.\n")
```

Finally, we return the full string back as the result.

```
        return (result)
    }
```

Using this function, we create Figure 9.1.

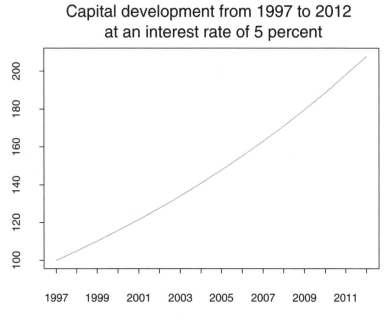

Figure 9.1. The development of a capital of $100 over 15 years, with an annual interest rate of 5%.

10. Useful Hints and Techniques

Now that you have seen elementary structures and techniques, as well as many advanced ones, we offer a few hints to make your work with S-Plus more efficient and enjoyable.

To start with, we give more detailed description about the internal functionality of S-Plus. This is followed by hints dealing with several aspects of programming and some tips and tricks from our own extensive experience with S-Plus. Since some of the hints we provide are based on our own experience, some people will find our ideas more useful than others. Nevertheless, we present several ideas about what might be helpful and what you could customize to have more fun with S-Plus and, of course, your data.

The following details are really only useful if you work with the system. For getting an overview, this section might be skipped, but coming back to it at a later point would be well worth the effort.

10.1 How S-PLUS works

To understand what S-Plus does during a session, we start by explaining some details about the startup and storing mechanism of the system.

10.1.1 Starting the S-PLUS System

.Data

When you start S-Plus, it searches in the directory from where it was started for a .Data directory (or _Data under DOS/Windows). The data directory is the place where all user created data and functions are stored.

If S-Plus finds this directory, .Data becomes the current working directory. If it does not find it, it searches for a .Data directory in the home directory (which is well defined under UNIX as the login directory, but needs to be set under DOS/Windows in the autoexec.bat or the splus.ini file). If there is such a directory, this becomes the working directory. If a .Data directory can't be found in either of these two locations, S-Plus creates it in the user's home directory and informs you about "Initializing a new S-PLUS user." All variables are now stored in this directory, unless explicitly specified to be stored in a different location. We will refer to this directory as the data directory in the following sections.

.First and .Last

After loading S-PLUS, the system searches for the .First function in the data directory. If it is found, S-PLUS executes this function immediately. When the session is ended, S-PLUS searches for a .Last function to execute, just as with the .First function at startup time. Various functions to customize your S-PLUS environment are normally stored in these functions. For example, many users like to open up the graphics window or attach some libraries on startup. For opening a graphics window, .First can be defined as follows.

Example 10.1. A .First function opening a graphics window on startup

```
.First <- function ()
{
  graphsheet()                     # start the device driver
  cat (".First ends here.\n")      # and print a message
}                                  # in the command window
```

Enter this definition and start S-PLUS again to see how (if) it works.

Storing data

When an expression is executed and no error occurs, S-PLUS stores the result of the expression in the assigned output variable, for example in y, if the statement is y <- f(x). This storing mechanism is a physical storing, which means that a file y is written to disk (to the data directory). The file is permanent, such that if you quit S-PLUS and start it again, the variable y will still be there, without reading anything in or re-entering the expression. The advantage of this mechanism is when you come back the next day, you can immediately pick up where you left off the day before. On the other hand, it requires some cleanup after a while.

S-PLUS has its own format of storing variables efficiently, so you cannot read the data files with a text editor. Even more, you should never copy anything into this directory unless you really know what you are doing. Most importantly, do not copy simple text files containing functions or data to this directory. Read them into S-PLUS using the appropriate functions, and S-PLUS will store them in the data directory with the correct format.

Under UNIX, objects are simply stored under the names you give them from within S-PLUS. This is not possible under DOS file systems (Windows 3.1, Windows 3.11, Windows95), because of DOS's limit of 8 (+3) character file names. If it is possible to store objects using their names, S-PLUS will do it. If not, it stores the objects under special names like __1 and creates a file ___nonfi (non–fitting names) containing a list of physical file names and corresponding object names in S-Plus.

.Last.value

Every expression entered in S-Plus modifies the variable .Last.value in the data directory. The result of the last expression is stored there, and if you enter an expression and forgot to assign the result to a variable, you can still retrieve the result from .Last.value. You must do this immediately, because even a command like objects(), which just lists all the variables in your directory, creates an output (the list of all variables) and stores its output in .Last.value.

⎡Note⎤ If an expression (a command input or a function called) does not end successfully, for example because of an error, variables are not modified at all. S-Plus restores the state of the system exactly to the state it had before the expression was entered.   ◁

10.1.2 Levels of Calls

In S-Plus, like in most modern computer languages, there are levels of calls. Each level keeps its own variables and must pass them explicitly to other levels. Only global variables are known to all routines. (In S-Plus, these are the variables of the interactive environment.)

Here is a short example designed to show the idea behind levels of calls. Try to figure out what value will be printed by the **print** statements if we enter **f(3)** after declaring the function as follows.

Example 10.2. The effects of levels in expressions
We declare a function g inside a function f and call the function with f(3). Think of what will be printed out with the statements print(x) we inserted.

```
f <- function (x)
{
  g <- function (x)
  {
    print (x)
    x <- x*x
    return (x)
  }
  print (x)
  store.x <- x
  x <- x*x
  print (x)
  y <- g(x)
```

```
    print (x)
    print (y)
}
```

Here is the output that will be shown if f is called with f(3):

```
3
9
9
9
81
```

What will happen if you enter g at the prompt now?
Is the value of store.x known inside the function g?

The variable store.x inside the function f, as well as the variable x inside g, which is passed to g from the calling function, are so-called *local variables*. They (or their values) are not known outside the functions in which they are declared. Such an environment, a closed function together with all values known within its scope, is called a *frame*.

| Note | If an expression is not completed successfully, nothing at all is stored. This is especially important if you intend to start a long simulation. If your function contains an error, which shows up after some minutes or even hours of runtime, no results are stored. So run a shorter version of such a simulation before you actually run the longer one, just to make sure everything works as expected. ◁

| Note | Good programming style avoids the use of global variables. For example, if your function needs a variable that is declared in the interactive environment, it will no longer work if the variable is assigned a new value or if you give this function to somebody else (without the external variable the function needs). (An exception might be setting default values to a generally used function, like an editor. These variables are typically "hidden" in UNIX convention; that is, their name starts with a period like .Last.value.)

This proper way of programming also ensures that you do not use a variable you didn't want to use because S-PLUS retrieved it from the global data directory. Accessing x without declaring it might still access a variable x in the interactive environment, probably unwanted. If the global x changes, the function might not work any longer. ◁

10.1.3 Searching for Objects

We learned from the previous section that S-PLUS first searches in the current data frame if a variable is referenced in a statement like x in 2*x. If no matching variable name can be found, the system searches through its search path. If working interactively, the search starts immediately there. The first entry in the search path is usually the current data directory. Afterwards, the system data directories are searched. If you attach a new directory by using the `attach` or `library` facility, these new locations are typically added after the data directory. In this way, you can replace system functions by simply "overwriting" them. If you assign something to a variable `hist`, for example, the variable `hist` is stored to your current data directory (or, to be exact, to the first element in your search path), and if you call `hist` afterwards, S-PLUS will find the new variable `hist` first. If you remove it, S-PLUS will again find the system `hist` function in the system directory.

To see the full search path list, enter `search()` at the prompt.

| Note | The S-PLUS system is a little intelligent while searching for variables. If a variable `print` is declared as `print <- 3`, and you then enter `print`, the value 3 is shown. However, if you enter `print (1/2)`, S-PLUS realizes that you want to execute a `print` function. You are then informed that another variable `print` was found, which is of mode numeric instead of being a function, and the function is executed (as you wanted).

◁

| Note | A confusing message occurs if a system function is "overwritten" in the way we just described. For example, variables of name c and `matrix` should be avoided, as system functions of the same name exist.   ◁

10.2 Tips for Programmers

There are certain techniques experienced programmers use as a standard. Most of them are easily discovered once you have worked for a while with the system. Nevertheless, knowing some of these in advance cannot do any harm.

10.2.1 Storing and Restoring Graphical Parameters

When writing a function around the creation of a graph file, you usually want to leave the function and restore the state of the graphics device to the state it was in before the function started. For this purpose, you can store all graphical settings into a variable to retrieve later. This is done by using (inside the function, before issuing any graphics command) the commands

```
> par.old <- par()          # save current graphics par.
> on.exit (par (par.old))    # and restore them on leaving
```

Note that putting the statement `par` (`par.old`) at the end of the function does almost the same thing, except that the status is modified if your function does not exit properly (crashes, for example). `on.exit` is always executed, whenever and however the function is left.

10.2.2 Naming of Variables and Functions

Some strategies become useful if you work with S-PLUS over a longer period of time. Objects (variables and functions) belonging together might become so numerous that their relationship fails to be visible when listing all objects in the data directory.

You typically create a lot of variables that you don't need in the long term. To identify these, it is a good idea to follow the rule that all temporary variables should start with x, y, or z. Using this rule, you can delete all these variables from time to time to clean up the data directory by entering

```
> remove (objects (pattern="x*"))
```

Variables and functions belonging together could start with the same string and have a period (.) and the actual identifier as the second part of the name. For example, if you examine your own car data, you might call the data itself `car.data`. The function to read the data from the external file could be named `car.read`, the function for plotting the data might be called `car.plot`, and the vector of means over the columns could be stored in `car.means`.

10.3 The Process of Developing a Function

If you develop a function, you absolutely must figure out how to set up a good environment. If you don't want to re-enter the function over and over again, you need to store it, edit it, and make the modifications permanent in S-PLUS.

One possibility of setting up a development environment is to start S-PLUS in one window and your favorite editor in another. You can work on the function and store it into a text (ASCII) file using the editor. The function can then be read into an S-PLUS session and executed by using

```
> source ("filename")
```

After examining the output, you might want to change the function and repeat the cycle.

Another possibility S-PLUS offers is to use its `fix` function. Enter

```
> fix (function-name)
```

to start up the editor (defaulting to a standard editor like vi or the Windows editor). Write the desired function and when you quit the editor, S-PLUS will automatically read in the new function. You may now execute the function as normal, and if something unexpected happens, you can rewrite the function by entering

```
> fix ()
```

without any argument. The default for fix is the last object "fixed." If there was a syntax error in your file (it happens to the best of us), and S-PLUS did not read it in successfully but gave an error message instead, you can still re-edit your input.

The advantage of the first approach is that you only need to switch between the editor and S-PLUS, whereas the second approach reads the new function in directly. So if you expect to do many changes (always double the number you expect), the first alternative with a separate editor might be preferable. This is especially true if you need to search for problems in the source code after receiving an error message from S-PLUS.

10.4 Imposing a Structure

If you want to work for a longer period of time with S-PLUS, the proposals in the following section might turn out to be useful to keep in mind, especially if there has been a pause in your work.

10.4.1 Working on Different Projects

If you start to work on several different S-PLUS projects, you may find it to be more of a hindrance than a help that S-PLUS stores all the objects you create in the same data directory. To avoid confusion, create several directories and have a separate data directory for every project. Under UNIX, simply create a directory named .Data in the location where you want to have the new data directory, for example in /users/michael/projects/thesis. Change to the desired location using the cd command and create the directory by entering

```
mkdir .Data
```

Under DOS/Windows, you need to do (almost) the same thing. Open a DOS Command Window, change to the desired directory, and create the directory _Data by entering

```
mkdir _Data
```

You then simply need to start S-PLUS from the directory that *contains* the data directory (not from the data directory itself). Under UNIX, switch to the directory and start S-PLUS from there. Under DOS/Windows, create a new icon (hold down the <Ctrl> key on the keyboard while clicking with the left mouse button on an S-PLUS icon and drag it to a new location). Edit the settings of the icon: In the [FILE] menu, click on [PROPERTIES] and change the entry *Working Directory* according to where you created the data directory.

To check if everything works as expected, start S-PLUS and enter the command `objects()`. This should show you an empty directory. If you then enter `search()`, your new data directory should appear in the first position.

10.4.2 Housekeeping-Cleaning Up Directories

After a while, the fact that S-PLUS stores all the objects you create in your data directory until you delete them (even after quitting a session) results in having lots of objects, some of which you don't need anymore. For this reason, we advise you to clean up the objects from time to time. The easiest way is to remove the unneeded objects one by one, using the `rm` command:

```
> rm (unneeded.object)
```

If you want to specify a set of objects, use the `remove` function. `remove` has many more possibilities than `rm`, but is slightly more complex to use. For example, you can delete objects in other data directories to which you have attached them (use `search()` to see them). The syntax for removing all objects beginning with an x from the second entry in your search path is

```
> remove (objects (pattern="x*"), where=2)
```

To remove absolutely all of the objects in the current directory (think before you try this out!), enter

```
> remove (objects())
```

If you are on a UNIX system, you can open a command shell and change to the data directory itself (for example, cd .Data). Then use the UNIX command

```
rm -i *
```

and the system deletes the files one by one, prompting you before deletion for every single file. *You may want to try this out in a test directory before actually going ahead and working on sensible data.*

10.5 Batch Jobs

Batch jobs are especially useful for multitasking systems. For example, if you want to run a simulation program overnight, or if you want to leave the terminal without staying logged on, you can run an S-PLUS program in the background using a batch technique.

A batch job consists of nothing more than a sequence of commands that the interpreter executes one after another. As this is a non-interactive technique, you do not need to intervene. The output is not written to the terminal, but to an output or log file.

There are only a few steps to do to run an S-Plus batch job.

- Create a simple text file containing S-Plus commands.
- If possible, check the code for correctness with the batch commands described below. For example, if you do a simulation with 10,000 iteration steps, run the program with 3 iterations to see if it works. If a syntax error occurs in a long simulation series, you will have wasted lots of time.
- Start the batch job by entering
      Splus BATCH inputfile outputfile
  from the command prompt (UNIX or DOS, respectively). The input file must exist, as it has to contain the commands. The output file is either created by S-Plus or overwritten if it already exists.
- Check if the program is running.
  On a UNIX system, enter `ps -ef` to see the processes.
  Under Windows, click on the background or press `<Alt><Tab>` several times to check the running processes, or press `<Ctrl><ESC>` to see the running processes.

Note You can produce and store graphs from a batch job in printer files, using the command `postscript` (or `win.printer()`). You cannot open a graphics window on the screen, as it will be closed after your batch job finishes. It may not open at all if somebody else has logged into the machine on which you are running the batch job (which cancels your permission to open a graphics device). However, you can send the graphics created to the printer and it will be printed out as usual.   ◁

10.6 Incorporating and Accessing C and Fortran Programs

The manual and other references explain quite well how to link a C or Fortran file to the S-Plus system. Therefore, we will restrict our explanation to a very short example. We will write a C function that takes a float parameter and gives back the third power (cube) of it.

Note S-Plus only exchanges pointers with external functions. For this reason, the self-written function in C or Subroutine in Fortran has to check the length of the data passed, as well as other possible sources of errors.
To pass the arguments properly, an explicit type conversion in S-Plus before calling the external function is adequate (e.g. `as.double` for double precision objects).   ◁

Example 10.3 shows the C function for calculating and returning the cube of a double precision value x.

Example 10.3. A simple C program

The following C function takes a double value x and returns its cube back to the calling routine. Note that the variables are always referenced by pointers.

```c
#include <stdio.h>
void cubic (double *x)
{ double *y;
  printf ("program cubic started.\n");
  *y = (*x)*(*x)*(*x);
  return (*y);
}
```

Example 10.4 shows a short function for passing the value x from the S-Plus environment to the C function and returning the cube of x back to the S-Plus environment. That is, after linking the function to the session, there is no visible difference between an internal and this external function call.

Example 10.4. The S-Plus function to interface the C module

The function cube.it in S-Plus calls the external C routine cubic, as defined above, and returns the result to the S-Plus environment. You will not realize the call is actually to an external C routine, unless you look at the function code.

```
cube.it <- function (x)
{
  the.cube <- .C ("cubic", as.double (x))
  return (the.cube)
}
```

Here are the essential steps to include (link) a C or Fortran program to S-Plus.

- Write a C program or a Fortran subroutine. Let us assume that the file's name is cubic.c, and the function's name is cubic. Do not declare a main() function!
- Compile the program to an object file.
  Typically, this is done with

  ```
  cc -c cubic.c
  ```

  for C functions. You should then have an object file cubic.o in your directory.
- Make the function known to S-Plus by loading it into the name table:

  ```
  > dyn.load ("cubic.o")
  ```

- Write the calling S-Plus function (see the example above).

That's it! You have already seen how to exchange data with the C program. You need to pass pointers and not the data itself. If you want to exchange vectors, for example, pass another parameter to the C function containing the length of the vector passed. For example, n can be integer and can be passed from S-PLUS like `n=length(x)`.

See the manuals for conversion rules from S-PLUS to C types (like in the example, where any numeric in S-PLUS is converted to double before being passed to C).

| Note | Note that using `dyn.load` links an object file to S-PLUS only for the current session. For permanent linking, you can link the object file to the system kernel (if you have the permission to do so), or do the linking in the `.First` function.   ◁

Windows users have to pay attention to some rules, as there is no standard C compiler. For further details, refer to the manual.

| Note | If you run into trouble using `dyn.load`, have a look at the function `dyn.load2` that comes with S-PLUS.   ◁

10.7 Exercises

Exercise 10.1

Create the directory Scourse and a corresponding data directory for S-Plus. Start S-Plus for the rest of this course from the new directory. Check if it works by starting S-Plus and determine what objects are already defined in this data directory.

Exercise 10.2

Start a batch job that does the following. Generate 1000 standard Normal random numbers. Store the random numbers in a variable. Calculate the mean and variance of the sample and print it out. Create a plot of the data and send it to your printer. Now start up S-Plus and interactively calculate the mean and variance of the random sample you generated by the batch job. Check if you obtain the same results.

Exercise 10.3

Write a function that creates a 2x2 layout for graphs, doubles the character size of text in the graph, creates four histograms of 100 newly generated standard Normal random numbers, and adds titles to the histograms. The number of samples to generate should be a parameter to the function, per default set to 100. Be sure that after the function exits, all graphics parameter settings are the same as they were before you started the function.
Verify this by creating another graph after the function is executed, and add a title to it.

10.8 Solutions

Solution to Exercise 10.1

Switch to the directory where you want to create your new project sub-directories, for example to /users/michael or C:\Splus\projects. Create the directory Scourse and the data directory by entering

```
mkdir Scourse
chdir Scourse
mkdir .Data  (for UNIX systems)
mkdir _Data  (for DOS systems)
```

Then start S-PLUS from here by entering Splus (on UNIX systems) or by creating a new S-PLUS icon and setting the working directory to, for example, C:\Splus\projects\Scourse (on DOS/Windows systems).

Solution to Exercise 10.2

When writing a batch job, we first need to write the batch job input. Use any editor you like to save the commands

```
x <- rnorm (1000)
print (mean (x))
print (var (x))
postscript ("graph1.ps")        # for postscript printers
win.printer ()                  # for Windows printers
hist (x)
title ("A histogram of 1000 random numbers")
dev.off()
```

to a file with the name, say *batch.in*. Use only one of the printer functions. Now you can run the batch job by entering

```
Splus BATCH batch.in batch.out
```

on the command line of your operating system.

After successful completion, you should have a file batch.out (and, if you used the postscript command, a file graph1.ps) in your directory. Look at the file batch.out and see what the mean and the variance of the sample were.
Open S-PLUS again and calculate the mean and the variance of x again. This should, of course, give the same result.
Note that the batch job created the new variable x in your directory.
If you used the command win.printer(), you should find a graphics page on your printer. If you used the postscript command, you can send the file to your postscript printer by entering

```
lpr graph1.ps
```

for UNIX systems or

```
print /b graph1.ps
```

for DOS systems.

Solution to Exercise 10.3

As shown before, using on.exit right at the beginning is the best method to write a function and leave the graphics parameters unmodified after returning from the function. Here is a short example.

```
do.the.graph <- function (n=100)
{
   par.old <- par()                 # store the old settings
   on.exit (par (par.old))          # activate on.exit
   par (mfrow=c(2, 2))
   par (cex=2)                      # set new parameters
   for (i in 1:4) {
      hist (rnorm (n),
         main=paste(n, "Normal random numbers"))
   }
}
```

The previous state is restored after the function exits, regardless of whether the function completed normally or crashed during its execution.
You can verify it by entering another command producing a graph, like

```
> hist (rnorm (1000), main="1000 random numbers this time")
```

You should see that the screen split into a 2x2 plot matrix is gone.
How about entering an invalid value for n in order to exit the function abnormally?

11. Special Topics

After having discussed some S-PLUS internal topics in the preceding chapter, we now discuss some practical hints and tips more related to S-PLUS and the "outer world," the hardware on which it runs and the software with which it cooperates.

We will discuss some aspects of implementing S-PLUS on a system using the library facility, which is especially useful in a multi-user environment. Later on, we show how graphics created in S-PLUS can be used with word processing systems. Finally, we provide references to electronic information resources.

11.1 Libraries

Installing a public library where functions of general interest can be stored often makes sense, especially on machines with multiple users. On single-user machines, creating a library is a good way of installing archives of functions for later access.

S-PLUS comes with some pre-installed libraries. If you enter

```
> search()
```

you see which paths are already appended to the standard directory. The output looks approximately like this:

```
[1] ".Data"
[2] "/usr/splus/splus/.Functions"
[3] "/usr/splus/stat/.Functions"
[4] "/usr/splus/s/.Functions"
[5] "/usr/splus/s/.Datasets"
[6] "/usr/splus/stat/.Datasets"
[7] "/usr/splus/splus/.Datasets"
[8] "/usr/splus/splus/library/trellis/.Data"
```

When you reference a variable, the directories are searched in this order. Only if the referenced variable is not found in any of these directories is an error message issued.

To add a new library to the current session, you enter the command

> `library` (*library-name*)

where *library-name* is either a directory in the library directory, or an absolute directory name like /users/mydir/Scourse/.Data.
To install a library, you can create a directory with S-PLUS data sets any-where on the system. Under the UNIX system, issue the command

`chmod 755 . *`

from the library directory to enable access to all users. To install a system-wide library, ask the system administrator to create a new library in the S-PLUS library directory, which is usually the directory library in the system path.
If you want to see what is in a specific library, enter

> `objects (where=6)`

to see all objects in the path listed on the sixth position of

> `search()`

To write something to the library, use

> `assign ("`*name*`", ` *data,* ` where=`*n*`)`

where *name* is the variable name (in quotes), *data* is the data (a valid S-PLUS expression), and *n* is the number of the search path list.

| Note | Under Windows, paths and directories are specified using backslashes (\\), as in C:\SPLUS. As the backslash is a special character in S-PLUS, it must be used with a preceding backslash, such that using the function `library` to access the directory C:\SPLUS\LIBRARY needs to be done in the following way.

> `library ("C:\\SPLUS\\LIBRARY")`

◁

11.2 Including Graphs in Text Processors

Usually a graph is created to be included in a text document. Among S-PLUS users, the typesetting system TEX (Knuth, 1991) and its enhanced system LATEX (Lamport, 1985) are very popular. S-PLUS graphs can easily be included in TEX and LATEX. In recent years, more and more S-PLUS users are working under the MS Windows system (in any of the variations Windows 3.1, 3.11, 95, and NT). The following section shows how to include graphs directly in Windows documents, and how to create and include PostScript files into text processing systems like Word and TEX.

11.2.1 Copying Graphs into Windows Text Processors

If a graph is created, it can be placed onto the Windows clipboard. For this purpose, click on the menu [FILE] and select the entry [SEND TO OTHER APPLICATION]. Switching to another application and clicking on [PASTE] in the [EDIT] menu inserts the graph at the current cursor position.

| Note | Windows users running an older version can create a graph on the clipboard by opening the clipboard as a device and proceeding as described above. This generates a graph of much better quality than simple Copy and Paste. An easy example of this is the following.

```
> win.printer(file="clipboard",format="placeable metafile")
> hist (rnorm (123))                # create a graph
> dev.off()                         # and close the device
```

◁

11.2.2 Using the PostScript Format

Possibly the best way to include S-PLUS graphics in other text documents is to create a PostScript graphics file and include it into the preferred text processor. You create a PostScript document by surrounding the actual graphics commands with commands that open and close the PostScript file:

```
> postscript ("filename.ps", height=5, width=6)
                                # opens a PostScript file
                                # of 6x5 inches
> graphics commands             # create a graph
> dev.off()                     # closes the PostScript file
```

For DOS/Windows based PCs, an alternative is to use the `win.printer` function, which uses your installed printer drivers. If you don't have a PostScript printer, you might nevertheless want to install a driver for it to be able to create PostScript graphs. In the following sections you will see why this can be useful. To save a graph from a graphics window of S-PLUS to a file, click on the [FILE] menu, then on [PRINT]. Activate the [PRINT TO FILE] check box before clicking [OK]. You are prompted to specify a file name before the graph gets saved and it will have a format that the active printer understands.

This file can now be imported in most standard word processors. We briefly mention some of the most common programs.

| Note | A PostScript file should not contain more than one graph or more than one page. A multiple page document is difficult to integrate into word

processors because of page layout issues and treatment of the graph as a floating object. ◁

11.2.3 PostScript Graphs in TeX

The combination of TeX or LaTeX as a text processor and S-Plus as a graphics system is very powerful. As we need to make a distinction between plain TeX and LaTeX (in both of its two variations, LaTeX 2.09 and LaTeX 2ε), we will treat them separately. We assume that a PostScript file is already generated and ready to be included into a text file.

Plain TeX. The plain TeX core system is actually not designed to include graphs by using a proper TeX command. TeX offers the "\special" command to include printer specific files. You need to consult your specific TeX installation manual to look up how to use the \special command.

A typical command to include a PostScript file is the following.

```
\midinsert
\special{psfile=filename}
\endinsert
```

LaTeX 2.09. For including PostScript graphs into LaTeX 2.09, we recommend that you use the tool EPSF by Thomas Rokicki, which can be found in most LaTeX distributions. The graphs in this book were created as PostScript files and included using EPSF.

If you do not have EPSF in your LaTeX version, you can retrieve it using ftp from the electronic archive LABREA.STANFORD.EDU in the US or FTP.DANTE.DE in Germany. These servers offer almost all TeX related material.

The EPSF documentation can be found at the beginning of the file. In its easiest variant, the command

```
\epsfbox{filename.ps}
```

includes the PostScript picture in the LaTeX document. It reserves the space it needs for the figure, but is only inserted when the TeX document is translated into a PostScript file (using dvips, for example).

Since including graphs is a very common problem, we list here the essentials of the LaTeX macro used to create this book (which allows automatic enumeration and referencing of figures from within LaTeX).

```
% --- The TeX Figure environment for including PS files ---
\newcommand{\Figure}[3]{%
{ % the macro has three parameters:
  % #1 - the figure undertitle
  % #2 - the figure index entry
  % #3 - the file name of the postscript graph
```

```
\begin{figure}
  \refstepcounter{figure}
  \centerline{\epsfbox{#3}}
  \noindent%
    {\small\bf Figure \thechapter.\arabic{figure}:} {#1}
\end{figure}
% finally write to the index file
\addcontentsline{lof}{figure}{%
  \hbox to 2em{\thesection\hfill} #2}
}
```

The macro has nothing special in it, it just takes the file name of the figure, the subtitle, and the entry to the lof (list of figures) file, which is typically shorter than the original undertitle. The figure appears centered on the page, and the text under the figure reads, for example, *Figure 3.2: A graph created in S-PLUS.*

| Note | EPSF allows you to scale a PostScript file within a LaTeX document, even after the graph is generated. Including the command

```
\epsfxsize=5in
\epsfysize=4in
```

scales the graph to be 5 inches wide and 4 inches high. If one of the two values is omitted, the other is adjusted proportionally.   ◁

LaTeX 2ε. For LaTeX 2ε, we can almost repeat what we have said for LaTeX 2.09. Note that a different package for the inclusion of PostScript files exists under the name EPSFIG.

If you don't know whether you are using LaTeX 2.09 or LaTeX 2ε, look at the beginning of one of your documents. A "\documentstyle" indicates that you are using LaTeX 2.09 (or LaTeX 2ε in its compatibility mode), whereas a "\documentclass" can only be used in LaTeX 2ε.

More TeX/S-PLUS Interfacing. There are many people working with TeX as a text processor and S-PLUS as statistics and graphics software. Due to this, there are many tools available interfacing TeX and S-PLUS. For example, on the statlib server (see Section 11.4, page 230), there is an S-PLUS function for writing S-PLUS data sets directly to a TeX table environment.

11.2.4 PostScript Graphs in MS Word

In MS Word (under Windows), there is the possibility of using the Copy and Paste function offered by Windows. Under S-PLUS, click in the graphics window to make it the active window, then select the entry [COPY GRAPH SHEET] in the [EDIT] Menu. Now switch to Word. In your document, click to indicate where you want to place the graph. Then select the entry [PASTE] in the [EDIT] menu and the graph is inserted into your document. This will always work, as it does not require any installed options from Word. If you print the document or zoom the graph by clicking on one of its edges and moving the mouse, you will notice that the quality of the picture is no longer as good as in the S-PLUS graphics window. The reason is that the Copy/Paste action copies bitmaps, that is, your picture is copied pixel by pixel. As your screen has a much lower resolution than the printer, printing with a higher resolution only doubles the pixels copied and the quality loss is literally visible.

A better way to insert the graph into your Word document is to use the [PASTE SPECIAL] menu entry in the [EDIT] menu. Using this feature, you can insert formats other than just text and bitmaps into your document.

You can try out the different methods, but if you have a PostScript printer, it might be the best idea to insert a raw PostScript file directly into your document. Click on [INSERT] and then on [PICTURE], then select your file and the file style, Encapsulated PostScript file. Word inserts the file into your document, and you can optionally select if you want the picture saved in the document or if Word should create a link to the file. The latter method updates the picture in your document if the original file is changed. On the other hand, if you delete the original file, it is no longer accessible to Word.

| Note | Word does not show the PostScript graph, even with the Preview option, but a number of lines of the PostScript code, like the title and creator's name. But if it is printed out, the pictures are properly inserted.   ◁

11.2.5 PostScript Graphs in Other Word Processors

For all other word processors, check to see if you can import PostScript files. If this is not the case, you should still be able to use S-PLUS's "Send to other application" facility and paste the graph into the word processing application.

11.2.6 If You Don't Have a PostScript Printer

Even if no PostScript printer is available, you can insert the figures as PostScript format files into your document. Create a PostScript file of your document (for TeX use dvips, for Windows programs install a PostScript printer, make it active, and redirect the printer output to a file) and then use a translation program to convert the PostScript file to a format suitable for your printer. The most common tool for translating PostScript format to other formats is GHOSTSCRIPT, available on most electronic archives.

The printout quality is very good and you can also preview the printout on the screen. About forty different devices are supported. After the conversion is done successfully, send the file created to your printer using `lpr` *filename*

on a UNIX system or

    `copy /b` *filename* `lpt1`

on a DOS/Windows system (check if lpt1 is your printer port).

11.3 S-News: Exchanging Information with Other Users

S-News is an E-mail (Electronic Mail) discussion list for S and S-PLUS users. An E-mail message is sent to a special address, and the server there sends the mail to all current subscribers of the discussion forum. The only thing you need is E-mail access.

11.3.1 Subscribing and Unsubscribing to S-News

You need to be aware of two addresses related to S-PLUS. If you want to send something to all subscribers, send your E-mail to

`s-news@utstat.toronto.edu`

If you want to get in touch with the administrator of the list to subscribe or unsubscribe, send a short request to

`s-news-request@utstat.toronto.edu`

Please *do not* send E-mail requests for subscription to the list address `s-news@utstat.toronto.edu`. Any mail sent to this address is automatically forwarded to thousands of other users, who will then flood you with replies like "Please send subscription request to ..." or "How could you be so stupid as to waste my time by ...".

    Your local site may have many subscribers to S-News. In that case, it may have a local mailing list, so that a single message from 'utstat.toronto.edu' is sent to the site and propagated to the local list. In this case, please contact your system administrator to be added to or dropped from the local S-News list.

11.3.2 Asking Questions on the Mailing List

Before you ask a question on the mailing list, check with your local S experts and read the FAQ (Frequently Asked Questions) file. (If you don't read the FAQ, you may inadvertently ask a "stupid" question and receive many responses from "holier than thou" S-PLUS *experts* saying, "If you had read the FAQ, I wouldn't be writing this E-mail message.") When answers are sent to you individually and not to the mailing list, it is considered good (n)etiquette to summarize the answers and mail them to the newsgroup. Your question should be as short as possible, reducing the problem to the core. It has better chances of being answered if it is of broad interest.

If you answer a question, pay attention who you send it to (for example, using the reply functionality of E-mail systems usually sends your answer to the whole list instead of just the originator of the message). If it is of general interest, send a cc (carbon copy) to the list.

11.4 The Statlib Server

The well-known statlib server is an archive for many topics in statistics. It also holds a large collection of S and S-PLUS related archives, functions, etc. The preferred way to use statlib is via the WWW (World Wide Web). For an explanation of electronic services, see for example, Krause (1995). The statlib home page is

`http://www.statlib.edu`

Statlib has ftp functionality under `lib.stat.cmu.edu`. The login is `statlib` and the password is your E-mail address.

Statlib can also be used via E-mail. Send the message

`send index`
    or
`send index from S`

to `statlib@lib.stat.cmu.edu` and nothing more, as the message is interpreted automatically.

11.5 R: A Public Domain Software

Two authors from New Zealand, R. Gentleman and R. Ihaka, wrote a program they call R. The name already suggests that the authors consider R to be a pre-stage of S. The system R has a syntax like S and covers the functionality as described in the *Blue Book* by Becker, Chambers, and Wilks.

R is available for free, including source code. It can run on hardware not supported by S and S-PLUS, especially on the Macintosh. It also runs on PC and UNIX systems and can be retrieved via anonymous ftp from

`stat.auckland.ac.nz`

12. References

12.1 Print References

Azzalini, A, and Bowman, A. W., 1990. A look at some data on the Old Faithful Geyser. Applied Statistics, 39: 357-65

Becker, R.A., and Chambers, J.M., 1984. S: An Interactive Environment for Data Analysis and Graphics. Wadsworth & Brooks Cole, CA

Becker, R.A., and Chambers, J.M., 1985. Extending the S System. Wadsworth & Brooks Cole, CA

Becker, R.A., Chambers, J.M., Wilks, A.R., 1988. The New S-Language. Wadsworth & Brooks Cole, CA

Becker, R.A., Cleveland, W.S., Shyu, M.-J., 1996. The Visual Design and Control of Trellis Display. Journal of Computational and Graphical Statistics, 5, 2: 123–155

Becker, R.A., 1994. A Brief History of S. In: Computational Statistics, Editors Dirschedl, P. and Ostermann, R.: 81–110. Physica, Heidelberg

Chambers, J.M., and Hastie, T.J., 1992. Statistical Models in S. Wadsworth & Brooks Cole, CA

Chambers, J.M., 1995. Overview of Version 4 of S. Technical report, AT & T Bell Laboratories, version of Jan 23, 1995. From AT&T electronic archive (see 'Electronic References').

Cleveland,W., 1993. Visualizing Data. Hobart Press, NJ

Devroye, L., 1986. Non–Uniform Random Variate Generation. Springer Verlag, New York

Everitt, B., 1994. A Handbook of Statistical Analyses using S–PLUS. Chapman & Hall, London

Härdle, W., 1991. Smoothing Techniques with Implementation in S. Springer Verlag, New York

Hilbe, J., 1996. Windows File Conversion Software. The American Statistician, 50, 3

Johnson, N.L., and Kotz, S., 1969a. Distributions in Statistics: Discrete Distributions. Wiley & Sons, New York

Johnson, N.L., and Kotz, S., 1969b. Distributions in Statistics: Continuous Univariate Distributions 1. Wiley & Sons, New York

Johnson, N.L., and Kotz, S., 1969c. Distributions in Statistics: Continuous Univariate Distributions 2. Wiley & Sons, New York

Johnson, N.L., and Kotz, S., 1969d. Distributions in Statistics: Continuous Multivariate Distributions. Wiley & Sons, New York

Knuth, D. E., 1991. Computers and Typesetting, Vol. A, The TeXbook, 11th ed. Addison-Wesley, Reading, MA

Knuth, D. E., 1991. Computers and Typesetting, Vol. B, TeX: The Program, 4th ed. Addison-Wesley, Reading, MA

Krause, A., 1995. Electronic Services in Statistics. Statistical Software Newsletter in Computational Statistics and Data Analysis, 19, 5: 595–604, (http://www.genedata.com/~andreas/internet/internet.html)

Lamport, L., 1985. LaTeX – A Document Preparation System. Addison-Wesley, Reading, MA

Marazzi, A., 1992. Algorithms, Routines and S Functions for Robust Statistics. Wadsworth & Brooks Cole, CA

Ripley, B.D., 1987. Stochastic Simulation. Wiley & Sons, New York

Sibuya, M., and Shibata, R., 1992. Data Analysis by using S. Kyouritsu Syuppan, Japan

Spector, P., 1994. An Introduction to S and S-PLUS. Duxbury Press, Belmont, California

S–PLUS, 1995. S–PLUS Documentation. Statistical Sciences, Inc.; Seattle, WA

Süselbeck, B., 1993. S und S-PLUS: Eine Einführung in Programmierung und Anwendung. Fischer, Stuttgart

The Economist, 1996. The Economist Pocket World in Figures. Profile Books, Ltd., London

Tukey, J.W., 1977. Exploratory Data Analysis. Addison-Wesley, Reading, MA

Venables, W.N., and Ripley, B.D., 1997. Modern Applied Statistics with S-PLUS, 2nd ed. Springer Verlag, New York

12.2 Electronic References

Electronic references can become outdated very fast. By the time of printing, a few references might already have changed. If you can't find information under the given address, you might try to find the new address or further information by using the popular search tools yahoo (http://www.yahoo.com) or lycos (http://www.lycos.com).

12.2.1 S–PLUS Related Sources

http://lib.stat.cmu.edu
The statlib archive containing a large S-PLUS archive and more.

http://netlib.att.com/cm/ms/departments/sia/doc/index.html
The AT&T report archive, containing (amongst others) many reports related to S-PLUS.

http://netlib.bell-labs.com/cm/ms/departments/sia/
    project/trellis/index.html
Further information about Trellis displays.

http://cm.bell-labs.com/cm/ms/departments/sia/jmc/index.html
John Chambers' home page referencing information on S.

http://www.stat.mat.ethz.ch/S-FAQ
Frequently Asked Questions collection on S and S-PLUS.

s-news-request@utstat.toronto.edu
Subscription address for the S–News mailing list. Personally managed.

s-news@utstat.toronto.edu
Address of the S–News mailing list. Mail is sent out directly to all subscribers.

12.2.2 TeX Related Sources

http://www.cdrom.com/pub/tex/ctan/CTAN.sites
ctan - Comprehensive TeX Archive Network. An overview giving references to different sites.

http://www.tex.ac.uk/
One of the ctan archives in the UK.

ftp.dante.de
TeX archive of the German TeX users group.

12.2.3 Other Sources

ftp.cs.wisc.edu:/pub/ghost/gnu/
GNU archive containing the Ghostscript PostScript tool.

Index